T0074058

5G, Cybersecurity and Privacy in Developing Countries

RIVER PUBLISHERS SERIES IN COMMUNICATIONS AND NETWORKING

Series Editors

ABBAS JAMALIPOUR
The University of Sydney,
Australia

MARINA RUGGIERI
University of Rome Tor Vergata,
Italy

The "River Publishers Series in Communications and Networking" is a series of comprehensive academic and professional books which focus on communication and network systems. Topics range from the theory and use of systems involving all terminals, computers, and information processors to wired and wireless networks and network layouts, protocols, architectures, and implementations. Also covered are developments stemming from new market demands in systems, products, and technologies such as personal communications services, multimedia systems, enterprise networks, and optical communications.

The series includes research monographs, edited volumes, handbooks and textbooks, providing professionals, researchers, educators, and advanced students in the field with an invaluable insight into the latest research and developments.

Topics included in this series include:-

- Multimedia systems;
- Network architecture;
- Optical communications;
- Personal communication services;
- Telecoms networks;
- Wifi network protocols.

For a list of other books in this series, visit www.riverpublishers.com

5G, Cybersecurity and Privacy in Developing Countries

Editors

Knud Erik Skouby
Aalborg University, Denmark

Prashant Dhotre
MIT School of Engineering,
MIT Arts, Design, and Technology University, Pune, India

Idongesit Williams
Aalborg University, Denmark

Kamal Kant Hiran
Sir Padampat Singhania University, India

River Publishers

Routledge
Taylor & Francis Group
LONDON AND NEW YORK

Published 2022 by River Publishers
River Publishers
Alsbjergvej 10, 9260 Gistrup, Denmark
www.riverpublishers.com

Distributed exclusively by Routledge
4 Park Square, Milton Park, Abingdon, Oxon OX14 4RN
605 Third Avenue, New York, NY 10017, USA

5G, Cybersecurity and Privacy in Developing Countries / Knud Erik Skouby, Prashant Dhotre, Idongesit Williams and Kamal Kant Hiran.

Routledge is an imprint of the Taylor & Francis Group, an informa business

ISBN 978-87-7022-647-9 (print)
ISBN 978-10-0084-803-8 (online)
ISBN 978-1-003-37466-4 (ebook master)

While every effort is made to provide dependable information, the publisher, authors, and editors cannot be held responsible for any errors or omissions.

Contents

Foreword

The WWRF Series in Mobile Telecommunications

The Wireless World Research Forum (WWRF) is a global organization bringing together researchers from industry and academia to identify the key research challenges and opportunities across a wide range of aspects of communication technologies. Members and meeting participants work together to present their research and develop white papers and other publications to guide us toward the Wireless World. Much more information on the Forum, and details of its publication program, is available on the WWRF website www.wwrf.ch. The scope of WWRF includes not just the study of novel radio technologies and the development of the core network, but also the way in which applications and services are developed, and the investigation of how to meet user needs and requirements.

WWRF's publication program includes use of social media, online publication via our website and special issues of well-respected journals. WWRF's own online journal will be launched soon. In addition, where we have identified significant deserving subjects, WWRF is keen to support the publication of extended expositions of our material in book form, either singly authored or bringing together contributions from a number of authors. This series, published by River Publications, is focused on treating important concepts in some depth and bringing them to a wide readership in a timely way. Some will be based on extending existing white papers, while others are based on the output from WWRF-sponsored events or from proposals from individual members.

Membership of WWRF brings considerable benefits for organizations and individuals who share our aims, including full involvement in our publication program. Further details are available on the website.

We believe that each volume of this series will be useful and informative to its readership, and will also contribute to further debate and contributions to WWRF and more widely.

Dr. Nigel Jefferies
Prof. Klaus David
WWRF Chairman
WWRF Publications Chair

Preface

As an emerging technology in mobile communication, 5G is expected to deliver a lot of improvements for industry and general users with respect to, e.g., new services and applications, geographical coverage and business possibilities. In particular, for specifically developing countries, 5G is seen as an important enabling factor for several of UN's Sustainable Development Goals where universal accessibility to ICT remains a serious concern. However, it is generally accepted that cybersecurity is a serious challenge, not exactly because of the broader usage and associated vulnerability. This also goes for developing countries that have additional challenges associated both with the expected faster build-up of accessibility and lack of qualified competencies within cybersecurity. Discussion of these challenges is the overall theme and motivation for this book.

The World Research Wireless Forum (WWRF) focuses on raising awareness and having discussions on the potential of 5G and beyond technologies, but also on the drivers and barriers related to implementation in different geographical regions of services and applications based on these wireless technologies. This book, the third in the series on ICT in developing countries, is motivated by ongoing activities, especially in Working Group A/B and the Cyber Security Working Group where the next generation technologies are discussed from technical, user and business perspectives. The issues are discussed both at a theoretical level and as cases in different countries across the continents.

Sincere thanks to authors:
Rakesh Kumar Birda, S J College, Jaipur, India; **Vineet Chouhan**, Sir Padampat Singhania University, Udaipur, Rajasthan, India; **Manish Dadhich**, Sir Padampat Singhania University, Udaipur, Rajasthan, India; **Ali Darvishi**, Allameh Tabataba'i University, Tehran, Iran; **Prashant Dhotre**, MIT School of Engineering, MIT Arts, Design, and Technology University, Pune, India; **Ruchi Doshi**, Universidad Azteca, Mexico; **Shubham Goswami**, Sir Padampat Singhania University, Udaipur, Rajasthan, India; **Darío M. Goussal**, School of Engineering, Northeast University at Resistencia (UNNE) - Resistencia,

Argentina; **Bhawna Hinger**, Govt. Meera (Girls) College, Udaipur, Rajasthan, India; **Kamal Kant Hiran**, Sir Padampat Singhania University, Udaipur, Rajasthan, India; **Tarannum Hussain**, Mohanlal Sukhadia University, Udaipur, Rajasthan, India; **Grace Dzifa Kwabena-Adade**, Graduate School, Ghana Communication Technology University; **Henoch Kwabena-Adade**, Accra Institute of Technology (AIT), Ghana; **Somayeh Labafi**, Iranian Research Institute for Information Science and Technology (IranDoc), Tehran, Iran; **Roslyn Layton**, Communication, Media and Information Technologies, Aalborg University, Denmark; **Mehul Mahrishi**, Swami Keshvanand Institute of Technology, Rajasthan, India; **Gaurav Meena**, Central University of Rajasthan, Ajmer, Rajasthan, India; **Hadi Moghadamzadeh**, University of Tehran, Tehran, Iran; **Robert Mwiinga**, Sir Padampat Singhania University, Udaipur, Rajasthan, India; **Shafi Pathan**, MIT School of Engineering, MIT Arts, Design, and Technology University, Pune, India; **Shalendra Singh Rao**, Mohanlal Sukhadia University, Udaipur. India; **Nilesh P. Sable**, Savitribai Phule Pune University, Pune, India; **Omega John Unogwu**, Centre for Geodesy and Geodynamics, National Space Research and Development Agency, Toro, Nigeria and Universidad Azteca, Mexico; **Ezer Osei Yeboah-Boateng**, National Communications Authority (NCA), Ghana; **Idongesit Williams**, Aalborg University, Denmark.

The editorial team is grateful for inspiration and support from the publishers and WWRF.

I, finally, but not least, thank Idongesit Williams, Kamal K. Hiran, and Prashant Dhotre for the tireless efforts spent in collecting the contributions and editing them.

The book addresses audiences within governments, organizations, industry, and universities.

Knud Erik Skouby,
May 2022
WWRF/Aalborg University

List of Contributors

Ali Darvishi, *Allameh Tabataba'i University, Iran*

Bhawna Hinger, *Govt. Meera (Girls) College, India*

Darío M. Goussal, *School of Engineering, Northeast University at Resistencia (UNNE) - Resistencia, Argentina*

Ezer Osei Yeboah-Boateng, *National Communications Authority (NCA), Ghana*

Grace Dzifa Kwabena-Adade, *Graduate School, Ghana Communication Technology University*

Gaurav Meena, *Central University of Rajasthan, India*

Hadi Moghadamzadeh, *University of Tehran, Iran*

Henoch Kwabena-Adade, *Accra Institute of Technology (AIT), Ghana*

Idongesit Williams, *Aalborg University, Denmark*

Kamal Kant Hiran, *Sir Padampat Singhania University, India*

Knud Erik Skouby, *Aalborg University Denmark*

Manish Dadhich, *Sir Padampat Singhania University, India*

Mehul Mahrishi, *Swami Keshvanand Institute of Technology, India*

Nilesh P. Sable, *VIIT, Savitribai Phule Pune University, India*

Omega John Unogwu, *Centre for Geodesy and Geodynamics, National Space Research and Development Agency, Nigeria and Universidad Azteca, Mexico*

Prashant Dhotre, *MIT School of Engineering, MIT Arts, Design, and Technology University, Pune, India*

Rakesh Kumar Birda, *S J College, India*

Robert Mwiinga, *Sir Padampat Singhania University, India*

Roslyn Layton, *Center for Communication, Media and Information Technologies, Aalborg University, Denmark*

Ruchi Doshi, *Universidad Azteca, Mexico*

Shafi Pathan, *MIT School of Engineering, MIT Arts, Design, and Technology University, Pune, India*

Shalendra Singh Rao, *Mohanlal Sukhadia University, India*

Somayeh Labafi, *Iranian Research Institute for Information Science and Technology (IranDoc), Iran*

Shubham Goswami, *Sir Padampat Singhania University, India*

Tarannum Hussain, *Mohanlal Sukhadia University, India*

Vineet Chouhan, *Sir Padampat Singhania University, India*

List of Figures

List of Tables

List of Abbreviations

2G	Second generation mobile
3G	Third generation mobile
3GPP	Third generation partnership project (3gpp)
4G	Fourth generation mobile
5G	Fifth generation mobile
5G-PPP	5G Infrastructure public private partnership
3Q	Third quarter
AI	Artificial intelligence
ACA	Argentine cooperatives association
AED	Automated external defibrillators
AF	Application
AMF	Access and mobility
ANN	Artificial neural network
API	Applications programming interfaces
AR	Augmented reality
ARPU	Average revenue per user
AUSF	Authentication server
AVM	Automated valuation models
BAP	Budget allocation process
CAPEX	Capital expenditure
CCPA	California consumer privacy act
CCTV	Close circuit television
CEA	Cybersecurity enhancement act
CI	Critical infrastructure
CIA	Confidentiality, integrity, and availability
CMM	Cybersecurity capacity maturity model for nations
DDoS	Distributed denial-of-service
DL	Deep learning
DMA	Digital market act
DSA	Digital service act

EB	ExaBytes
EC	European commission
EMS	Emergency medical services
EOS	End of support
EU	European union
EULA	End user license agreement
FAO	Food and agriculture organization
FCC	Federal communication commission
FFA	Financial fraud action
FWA	Fixed wireless access
GAO	US Government Accountability Office
CAGR	Compound annual growth rate
GCA	Global cybersecurity agenda
GCI	Global cybersecurity index
GCSCC	Global cyber security capacity centre
GDP	Gross domestic product
GDPR	General data protection regulation
GHz	Gigahertz
GPS	Global positioning system
GSMA	GSM Association
GNSS	Global navigation satellite system
GWN	Gateway nodes
HaaS	Hardware as a service
ICT	Information and Communications Technology
IDB	Inter-American Development Bank
IMS	Ip multi-media subsystem
IMSI	International mobile subscriber identity
InSAR	Interferometric synthetic aperture radar
INTA	Argentina´s National Agricultural Technology Institute
IoT	Internet of things
IP	Internet protocol
ITU	International telecommunications union
Kbps	Kilobits per second
LGPD	Lei geral de proteção de dados
LoRa	Long range
LP-WAN	Low power wideband area networks
LR	Logistic regression

LTE	Long term evolution
LTE-M	Long term evolution for machines
M2M	Machine-to-machine communications
Mbps	Megabits per second
MDT	Mobile data terminals
MEC	Multi-access edge computing
MHz	Megahertz
MITM	Man-in-the-middle
MIoT	Mobile internet of things
ML	Machine learning
MMIMO	Massive MIMO
mMTC	Massive machine type communication
MMW	Millimeter wave
MNO	Mobile network operator
MVNO	Mobile virtual network operator
NAS	Non-access stratum
NB-IoT	Narrowband internet of things
NCA	National communications authority
NDAA	National defense authorization act
NDVI	Normalized difference vegetation index
NFV	Network function virtualization
NGO	Non-governmental organizations
NIST	National institute of standards and technology
NPN	Non-public networks
NR	New radio
NS	Network slicing
NSA	Non-standalone architecture
NSSF	Network slice selection
NTIA	National Telecommunications Information Agency
NWDAF	Network data analytics function
O-RAN	OpenRAN
OAS	Organization of American States
OPEX	Operational expenditure
OTT	Over-the-top
PDA	Personal digital assistance
PCF	Policy control function
PKI	Public key infrastructure

PN	Public network
POPIA	Protection of personal information act
RADAR	Radio detection and ranging
RAN	Radio access network
RIC	RAN intelligent controller
RFID	Radio frequency identification
SaaS	Software as a service
SAR	Synthetic aperture radar
SaMD	Software as a medical device
SBA	Services based architecture
SCD	Social and cultural dimension
SCP	Service communication proxy
SDN	Software defined networking
SEPP	Security edge protection proxy
SIM	Subscriber identification module
SMF	Session management function
SUCI	Subscription concealed identifier
SUPI	Subscription permanent identifier
UHF	Ultra-high frequency
UPF	User plane function
URLLC	Ultra-reliable and low-latency communication
USD	US dollar
UWB	Ultra-wideband
VoLTE	Voice over lte
VR	Virtual reality
VTF	Vertical engagement task force
Wi-Fi	Wireless fidelity
WoM	Word of mouth
WRC	World radio communication conference
WSN	Wireless sensor networks

Introduction

Wireless Networks and Privacy in Developing Countries

Knud Erik Skouby, Idongesit Williams, Prashant Dhotre, and Kamal Kant Hiran

Background of the Book

Wireless networks have shown to be the fastest-growing communications technology in history. Wireless networks such as mobile telephony emerged commercially in the 20th century as a complimentary service to fixed-line telephony. However, the availability, affordability, and accessibility of mobile cellular technologies in the 1990s upwards changed the telephony landscape. Some of the drivers of these changes included regulatory reforms in the telecoms sector aimed at promoting a competitive market, innovative business modeling, innovative service offering, and most importantly demand. Developing countries, though laggards in the adoption of previous generation of mobile telephony, were not left out of the mobile revolution. By the 2000s, developing countries, prompted by external market forces and the need to modernize their telecommunications sectors joined the bandwagon. Furthermore, developing countries such as Kenya saw the opportunity of delivering value-added services such as mobile money to expand their mobile networks. One cannot also write off the impact of globalization in service delivery, socialization, and trade as drivers that sustained the growth in the uptake of mobile broadband and mobile internet services.

Over the years, the evidence of the deep connection between wireless communication and development has been analysed, confirmed in research, and documented in reports from, e.g., OECD and ITU. Wireless communications have supported ICT for development initiatives in developing countries. There are examples of some of these initiatives in developing countries. One example is that of Esoko Ghana focusing on e-agriculture using mobile

1

telephony (Ghana). Other examples include Fundapi Ecuador focusing on the delivery of technical capacity building using social media, and Tula Salud Guatemala focusing on rural e-health to name a few.

The result of the regulatory drivers, market drivers, and upshot of development initiatives using mobile, has been the rapidly falling costs and technological progress. The outcome of this result, evident several years ago is the fact that the number of mobile subscribers has surpassed the number of fixed subscribers. Furthermore, it has made connectivity to rural and remote areas feasible. Hence, as indicated earlier, 95% of the world population now has access to a mobile broadband network. However, a coverage gap remains significant in Africa, where 18% of the population remains without any access to a mobile broadband network in 2021. The most recent wireless network, 5G is expected to address this by connecting people, things, data, applications, transport systems, and cities in smart networked communication environments. The business cases and scope of use present challenges, and these new functionalities and new services necessitate a new way of deploying advanced mobile services.

There appears general agreement, that the impact in developing countries will especially emerge through, e.g.,

- Smart transportation systems: According to WHO, 90% of the world's fatalities on the roads occur in low- and middle-income countries, even though these countries only have approximately half of the world's vehicles.

- e-Health: Remote surgery will reduce the latency to enable remotely assisted surgery. Specialists are not available in many hospitals and could join a local surgeon remotely to perform procedures that require expert skills (5G's latency will be around one millisecond – unperceivable to a human and about 50 times faster than 4G).

Apart from business case challenges that are not discussed in the present volume, also issues related to technologies and security/privacy remain to be solved and aspects of these are discussed. The value of the discussion on cybersecurity and privacy is based on the fact that 5G will support an increase in connected devices. As the connectivity density increases, so will there be an increase in Internet traffic as well as cybersecurity and data privacy incidents (Kox, 2013). These incidents all have cost implications to the mobile network operator and the overall economy at large (ibid). Developing countries, as will be discussed in this book, are not immune to these challenges – on the contrary. Cybersecurity competencies are in short supply and high demand globally and this is likely to represent a serious challenge to developing countries. Hence the idea behind this book was conceived to explore existing and potential cybersecurity and data privacy challenges and solutions with respect to 5G and what should be done about these challenges. In the introductory section of the book, the background of the book is followed by the state of the art on 5G cybersecurity and privacy in developing countries. This is followed by a description of the book and the target audience of the book.

5G, Cybersecurity, and Privacy in Developing Countries

5G was launched commercially in 2019. Three years down the line, 5G has been deployed in a sizable number of developing countries in every region around the globe. There are also a lot of upcoming deployment initiatives taking place in developing countries around the globe. Public regulatory authorities and network operators in some developing countries are taking the first steps by providing the 5G infrastructure that will drive the aforementioned application areas. In the last three years, network operators and government authorities have been laying down investment plans and spectrum allocation plans respectively in countries in Africa (GSMA Intelligence, 2021), the Middle East, and the Americas. As of 2021, 5G networks are launched in South Africa, Togo, Argentina, Brazil, Madagascar, Peru, Columbia, Saudi Arabia, Malaysia, Indonesia, Philippines, China, Laos, Uzbekistan, and Tajikistan. It was being deployed in Chile, Libya, Kenya, Swaziland, Uruguay, Ecuador, and Sri Lanka. There are investments in Nigeria, Cameroon, Gabon, Republic of Congo, Angola, Uganda, Egypt, Algeria, Morocco, Western Sahara, Papua New Guinea, India, Pakistan, Mongolia, Nepal, Bangladesh, Cambodia, Bolivia, Mexico, and Costa Rica (ibid). This indicates that despite the challenges with the business cases with respect to the implementation of 5G use-cases, mobile network operators have identified alternative business cases and use cases for the development of 5G in developing countries.

Hence the dream of 5G in developing countries is from a technical perspective no more in the future but right here. 5G is expected to build on the technical value already delivered by 4G to organizations, individual lives, and society as large. It is also expected to build tremendously on the positive network effects currently provided by 4G.

The launch of 5G opens the possibility for an increase in connected devices in developing countries in the years to come. According to CIO 5G can support "1 million connected devices per 0.38 square miles" compared to 4G. 4G is estimated to support "2000 connected devices per square miles" (ibid). This opens up the possibility for the implementation of smart manufacturing (industry 4.0), massive IoT networks in different verticals, and the delivery of data-intensive services using virtual reality, artificial intelligence, augmented reality, etc. Although it is hard to predict at the moment, there are signs that these services and their respective markets will exist in some developing countries. As examples, there are informal structures in developing countries that work to promote the development and uptake of smart manufacturing in Africa. An example includes partnership for the 4th Industrial Revolution in South Africa (4IRSA). There are AI-driven and IOT services respectively in every region in developing countries around the world. Examples include AI service providers such as Zensar (India), Flavoursoft (Nigeria), AI Kenya (Kenya), S4 (Argentina), Arara (Chile), Shelfpix (Brazil), Mozn (Saudi Arabia) to mention a few. Examples of companies providing IoT services in developing countries include IoT solutions (South Africa), Xperanti (Malaysia), Eurisko mobility (Lebanon), Elera IoT network in Latin America to mention a few. Based on these examples among others out there, it is likely that developing countries will over time also experience an increase in connected devices in tandem with the rate of the rollout of new innovative services based on these technologies. We can see evidence of this possibility with the introduction and growth of mobile money services in developing countries over the years.

It should be noted that the current push by mobile network operators to adopt 5G in developing countries is not based only on the promise of connected devices. This is because, currently, the level of adoption of services that will result in an exponential increase in connected devices (such as via IoT) is low in developing countries. Reasons for the low availability of these services range from the lack of an enabling operational environment and use cases on the supply side to the lack of demand can be attributed to the low level of adoption of these services. The promise rather stems from the increasing rate of subscription and consumption for mobile Internet services in developing countries as mentioned earlier. ITU records that there was an active growth in broadband subscriptions of 18.4% a year between 2005 and 2019. In 2019, the number of mobile internet subscriptions in developing countries stood at 74 subscriptions per 100 inhabitants. There was a slight decline in subscriptions in 2020, probably due to the COVID-19 pandemic. However, in 2020, 72% and 34% of the inhabitants of developing countries in urban and rural areas respectively were active internet users (ibid). The current level of adoption of mobile internet services correlates with the level of penetration of 3G and 4G coverage in developing countries. ITU records an 85% and 95% coverage of 4G and 3G networks respectively, as of 2021 (ibid). Hence there is an opportunity for MNOs to acquire and maintain more active internet users and active mobile subscribers in developing countries. This is especially so in sub-Saharan Africa are rural areas in the Arab states and the Asia Pacific region where the level of internet usage is low. This promise of increased subscription and consumption of mobile Internet services in developing countries makes it relevant to consider the implementation of 5G in developing countries. This promise potentially grants MNOs, who are early birds in the acquisition of 5G spectrum, competitive advantage over other MNOs as they can provide subscribers not only with better access to the Internet but also with a broader range of- service offering that caters for entrepreneurial initiatives that require 5G networks.

Hence mobile network providers, building on the existing and potential demand for data-driven services in developing countries, have not hesitated in pushing national authorities in developing countries for spectrum to deliver 5G (GSMA Intelligence, 2021). The necessary investments in 5G networks seem to be an even bigger challenge than for 4G; these challenges are, however outside of the scope of the discussion here. Public regulatory authorities, as mentioned earlier, in developing countries have nevertheless seen the need for data-driven services in developing countries. Existing data-driven services have given rise to the possibility of e-governance. It has also resulted in the emergence of innovative service delivery which has resulted in new employment possibilities and a more efficient approach to both public and private service delivery. As a result of this potential, some public regulatory authorities in developing countries are allocating and authorizing the use of spectrum for 5G (GSMA Intelligence, 2021). These factors among others are some of the reasons for the early deployment of 5G in developing countries.

However, the growth and impact of wireless network in developing countries has been accompanied with increasing data privacy and cybersecurity challenges. This is evident in the current cybersecurity challenges that affect mobile networks and mobile devices in different sectors in developing countries supported and enabled by mobile wireless networks. The cybersecurity incidents recorded in developing countries are not limited to cyberattacks using botnets, infostealers, ransomware, phishing, denial of service, cryptojacking, malware attack, account takeover, etc. There are reports of the extent of these cybersecurity challenges at a regional level and nationally. As an example, at a regional level, it is assessed that in Africa, 61% of companies were affected by ransomware in 2020. Nationally, In Brazil, the reported cases of cybercrime grew from less than 4000 cases in 1999 to more than a million cases in 2014. Approximately 76% of these attacks originated from Brazil. The incidents included denial of service, worm, scam, cyber fraud, and invasion. Another example is Vietnam. In Vietnam, ransomware is said to have increased by nearly 200% compared in 2020 as compared to 2021.

Although evidence of cyberattacks on the core infrastructure of wireless networks in developing countries is not readily available, it cannot be ruled out. Such attacks have been recorded in developed countries. For example, recently the servers of Vodafone Portugal's 4G and 5G networks experienced a cyberattack. As wireless networks continue to evolve as critical infrastructure, so will it become attractive for cybercriminals who could be either after sensitive data in the database, impersonation or holding the network as ransom for whatever reason? Such reasons could include state-sponsored espionage or a cybercriminal driven by either economic or ideological reasons.

One of the consequences of cyberattacks is data breaches. The data breach becomes serious if it affects sensitive data, such as personal data. The challenge though is that developing countries have encountered their own fair share of data breaches. In some cases, the amount of data breaches annually is alarming. For example, in 2013, about 10 million, 1 million, 700,000, and 30,000 data records were lost in Pakistan, South Africa, Vietnam, and Chile respectively. It is not clear if these data breach is from either local or cross-border actors. What is also not often clear is how developing countries handle these incidents.

These cybersecurity incidents are not without economic consequences. For example, South Africa loses $157 Million annually to cybercrime. In 2019, Nigeria lost 200 billion Naira (Approx $482 million) annually to cybercrime. Furthermore, between 2015 and 2017, Indian banks lost RS 88553 ($1152) an hour to cybercrime. In developing countries, these crimes are motivated by poverty and the attackers embark on cybercrime due to economic motivations.

Although the security features of 5G are enhanced, there is actually no guarantee that it will be free of cyberattacks. Furthermore, the activities of the emerging economy and society – not only in developing countries – will be dependent on the functional operation of 5G (and beyond) networks. Hence a cybersecurity breach could result in catastrophe or even fatality in the case of self-driving cars, a platoon of logistic trucks or even the destabilization of economic and social activities.

As 5G is being deployed, there is a need for discussion on ex-ante regulatory and operational measures that will result in a secure 5G environment in developing countries. These are measures that support different stakeholders in the 5G ecosystem. As a result, developing countries will be continuing to benefit from secured wireless network. This book provides analysis, reflections, and empirical studies with proposals on what those measures should be. In the next section, a description of the book and audience of the book is provided.

About the Book

The book is divided into four parts with nine chapters.

Part 1

This part consists of three chapters, chapters 1, 2, and 3. The focus of the first part of the book is on what developing countries should consider when developing national cybersecurity regulations for 5G. There are discussions on privacy within the context cybersecurity in this part of the book. Although it is not always stated explicitly, the chapters in this part of the book provide inputs into missing pieces of national cybersecurity policies in developing countries. The three chapters view national regulatory frameworks as important because these frameworks will help in shaping the cybersecurity landscape in developing countries. There is an agreement in the three chapters on the following points:

- Current cybersecurity challenges will manifest itself in 5G networks globally as well as in developing countries.

- There are potentials for increased cybersecurity threat with 5G compared to what exists now in developing countries hence developing countries require national cybersecurity policies.

- A technical, economic, and operational mindset is required for the development of these dedicated 5G and beyond cybersecurity policies.

As a result, there is a need for update of current and development of new cybersecurity policies to cater for the existing threats. However, the proposals and arguments ,in the three chapters, on their proposals, on the aforementioned three points (potential policy objectives), policies differ.

In Chapter 1, there is a proposal for the consideration of the cybersecurity and privacy needs for rural areas in such policies. Often cybersecurity policies do not demarcate cyber threats along socially defined geographical boundaries. This often is because the same wireless network and similar services are made available to all. Hence the threats are uniform. However, with 5G, there will be demarcations in verticals deployed in urban areas in some developing countries and those deployed in rural areas. One does not expect to see self-driving cars, even with the launch of 5G in the near future in rural areas in developing countries, neither does one expect to see large-scale smart farming in urban areas. But as mentioned in this book, one is likely to find 5G-dependent services supporting e-agriculture, e-health, etc., in rural areas. Hence there will be cybersecurity and privacy needs in developing countries for 5G-enabled services in rural areas. Currently, these needs in current wireless networks in rural areas are unknown. The chapter argues that this should not be the case with 5G. Therefore, it is imperative that such needs be identified, and policies drafted to cater for these needs. This chapter provides demonstrable framework for Latin America, using e-agriculture and e-health as cases on how to identify these needs. The chapter provides an insight on what the 5G technical, economic, and operational mindset when considering 5G rural cybersecurity policies should be in Latin America and developing countries at large.

Chapter 2 proposes a holistic 5G ecosystem approach to regulating cybersecurity in 5G networks. The chapter stops short of proposing cross-border regulations that would have an impact on the 5G supply chain. The challenge though is that mostly very few suppliers exist in developing countries. Hence the chapter draws a national boundary around the specifics to be considered in such regulations.

Chapter 3 argues for the prioritization of crises management and recovery in 5G cybersecurity regulations. The case analyzed in this chapter is India. The argument for the prioritization of crises management and recovery is that 5G will indeed be a critical infrastructure in India. The chapter provides inspirational risk management framework looking at technical operational and behavioral issues that could be a part of 5G cybersecurity regulations.

The adoption of these proposals into national cybersecurity policies in developing countries opens up the possibility for protecting the data and privacy of rural subscribers, ensure that MNOs and application providers have a national guide for managing cybersecurity risks; ensure that policymakers understand areas to consider when either drafting new or upgrading existing cybersecurity laws.

Part 2

The second part of the book provides insight into potential cybersecurity challenges with regards to 5G supply and service delivery and have two chapters, chapters 4 and 5. The focus of chapter 4 is to understand the general security implications of OpenRAN compared to conventional RAN. This chapter is important because the deployment of telecom infrastructure in rural areas in general has always been a problem for mobile network operators. This is due to the lack of economic viability of the telecom services in rural areas. However, as mentioned earlier there is a rapid deployment of 5G in developing countries. This has been made possible in most cases using OpenRAN. According to Deloitte, 60% of 5G deployments and trials in rural areas in developing countries are with OpenRAN. However, how valuable is

OpenRAN to developing countries from a cybersecurity perspective? That is the quest behind this chapter. The chapter explores different arguments for and against OpenRAN compared to regular RAN in developing countries. Although the chapter does not provide direct support for one or the other, it provides food for thought for MNOs and policymakers actively deciding on the option involving a greater security challenge when deploying 5G. Chapter 5 assesses the potential cybersecurity challenges to the adoption of 5G network and services in Ghana. 5G is yet to be launched in Ghana; however, it is expected to support eLearning, security services, road safety, elections, etc. However, there are current cybersecurity challenges that will also affect Ghanaian users when they adopt 5G for AI, Machine learning, and IoT to support the aforementioned services. The chapter highlights the need for regulation but points to the greater need for user e-Identification as a means of dealing with the unique cybersecurity challenges in Ghana.

Part 3

The focus of this part is on data privacy. It consists of Chapters 6 and 7. Chapter 6 is a case study on the legal framework for data privacy in Iran. As mentioned earlier, most developing countries possess data privacy laws. However, the governance approach to these laws varies. In the case of the GDPR, there is a centralized governance approach, and the roles of the interconnected actors are clear. That is not always the case. In some developing countries such as Lebanon as an example, the governance approach is decentralized where a data controller can process personal data with permission from the ministry of economy and trade. Otherwise, it requires direct consent from the user (ibid). In the absence of stringent enforcement mechanisms, the data controller can bypass the law and hope not to be caught by the data subject. This can also be the case as it pertains to data processing by a data controller delivering 5G services. Hence chapter 6 presents an analysis of the data privacy regulatory framework and the actors involved in Iran. The chapter highlights the problems in the regulatory framework and proposes measures on how they can be fixed. This is a chapter that will also be of value to policymakers. Chapter 7 laws present an empirically driven chapter on how data privacy, among other factors, will potentially affect the consumption of 5G services in developing countries.

Part 4

The fourth part of the book provides 5G-related application cases where issues on cybersecurity are dealt with. There are two chapters in this section,

Chapters 8 and 9. Chapter 8 describes how machine learning can be used to predict and detect threats during financial transactions in 5G networks and services. Chapter 9 describes how interferometric synthetic aperture radar, global navigation satellite system, 5G, IoT, and cybersecurity can be integrated for earthquakes and deformation monitoring.

The Target Audience of the Book

Different authors from South America, Asia, Middle East, Africa, and Europe have provided their thoughts on the questions raised in the book. The authors are mostly academics and consultants. The writing style is a mix of academic and consultancy approach. The book consists of exploratory studies, case studies, and descriptive analysis. The target of the book is not limited to the academia, knowledge institutions, public authorities, think tanks, industry players, and nongovernmental institutions. Nevertheless, the book is also written in a way that anyone with interest in the topic can read it.

References

Lewis, M. (2021). *Telstra pushing 5G into regional and rural Australian communities*. Retrieved from Mobilecorp, Sydney-Australia: https://www.mobilecorp.com.au/blog/telstra-pushing-5g-into-regional-and-rural-australian-communities

Ahmed, A. (2021). Lightweight digital certificate management and efficacious symmetric cryptographic mechanism over industrial Internet of Things. *Sensors*, 2810.

Akinsolu, M., Sangodoyin, A., & Adeyemi, K. (2021). Design considerations and data communication architecture for national animal identification and traceability system in Nigeria. *IST-Africa 2021 Conference Proceedings, IST-Africa Institute/ IIMC.*

Ali, R., Pal, A., Kumari, S., Karuppiah, M., & Conti, M. (2017). A secure user authentication and key-agreement scheme using wireless sensor networks for agriculture monitoring. *Future Generation Computer Systems, 84*, 200 - 2015.

Aranda, J., Sacoto-Cabrera, E., Haro-Mendoza, D., & Astudillo, F. (2021). 5G networks: A review from the perspectives of architecture, business models, cybersecurity, and research developments. *Novasinergia, 4*(1), 6–41.

Arizton Research. (2022). *Latin America Data Center Market Report Scope 2021–2026*. Arizton Research.

Beavers, I. (2018). Intelligence at the edge Part 3: Edge node communication. *Technical Report, 2018. Analog Devices Engineer Zone.*

Bestsennyy, O., Gilbert, G., Harris, A., & Rost, J. (2021). *Telehealth: A quarter-trillion-dollar post-COVID-19 reality?* McKinsey Insights.

Biamis, A., & Curran, K. (n.d.). 5G security and the Internet of Things. In *Security and Organization within IoT and Smart Cities.* CRC Press: USA.

Bronson, K. (2019). Looking through a responsible innovation lens at uneven engagements with digital farming. *NJAS-Wageningen Journal of Life Sciences*, 90 - 91.

Brummer, T. (2021). Cybersecurity in beyond 5G: use cases, current approaches, trends, and challenges. *Communication Systems XIV, Technical Report IFI-2021.02 Chapter 3.*

Cabral, E., Silva Castro, W., Florentino, D., Araujo Vianna, D., daCosta Junior, J., Pires de Souza, R., et al. (2018). Response time in the emergency services. Systematic review. *Acta Cir. Bras., 33*(12), 1110–1121.

Campbell, C. (2019). *What the Chinese Surveillance State Means for the Rest of the World.* Retrieved October 19, 2021, from https://time.com/5735411/china-surveillance-privacy-issues/

Casey, K. (2021). *Edge computing and IoT: How they fit together.* Retrieved October 19, 2021, from https://enterprisersproject.com/article/2021/3/how-edge-computing-and-iot-fit-together

CEA. (2014). Cybersecurity Enhancement Act of 2014 (Enacted January 1, 2021). *Public Law*, 113–274.

Chow, W. (2021). *The global economic impact of 5G" Global Technology, Media and Telecommunications (TMT).* PWC.

Cordeiro, M. (2021, 128). *Brasil ganha primeira antena 5G voltada para o agro no sul do país.* Retrieved from Digital Policy Law: https://digitalpolicylaw.com/sercomtel-realizara-proyecto-piloto-5g-en-areas-rurales-de-brasil/

Correa Lima, G., Lira Figueiredo, F., Barbieri, A., & Seki, J. (2020). Agro 4.0: enabling agriculture digital transformation through IoT. *Revista Ciência Agronômica -.*

Cosby, A., Manning, J., Fogarty, E., & Wilson, C. I. (2021). *Assessing real time tracking technologies to integrate with identification methods and national traceability requirements.* North Sydney: Final Report, Project V.RDA.2005. CQ University Australia. Meat and Livestock Australia Ltd.

Creese, S., Dutton, W., & Esteve-González, P. (2021). The social and cultural shaping of cybersecurity capacity building: a comparative study of nations and regions. *Personal and Ubiquitous Computing*, 1 –15.

CSRIC 7. (2020). *Report on recommendations for identifying optional security features that can diminish the effectiveness of 5g security - Communications Security, Reliability and Interoperability.* USA: Working Group 3: Managing Security Risk in Emerging 5G Implementations.

Davies, J., & Goldberg, R. (1957). *A concept of Agribusiness.* Boston USA: Division of Research, Graduate School of Business Administration, Harvard University.

Dimitrievski, A., Filiposka, S., Melero, F. J., Zdravevsvki, E., Lameski, P., Pires, L. M., et al. (2021). Rural healthcare IoT architecture based on low-energy LoRa. *International Journal of Environment Research and Public Health, 18*(2021), 7660.

Dorairaju, G. (2021). *Cyber security in modern agriculture case study: IoT-based insect pest trap system.* MsC. Thesis, Jamk University of Applied Sciences, Finland.

Drougkas, A., Liveri, D., Zisi, A., & Kyranoudi, P. (2020). *Procurement guidelines for cybersecurity in hospitals.* European Union Agency for Cybersecurity (ENISA).

Duffy, C. (2020). *The big differences between 4G and 5G.* Retrieved October 19, 2021, from https://edition.cnn.com/2020/01/17/tech/5g-technical-explainer/index.html

Electroral Commission Ghana. (2021). *2020 Presidential Election Update.* Retrieved October 1, 2021, from https://ec.gov.gh/

Ericsson. (2018). *A guide to 5G network security.* Insight Report 18:000589 Uen Ericsson AB.

Ericsson. (2021). *Edge computing - a must for 5G success.* Retrieved October 1, 2021, from https://www.ericsson.com/en/edge-computing

Ericsson. (2021). *Ericsson Mobility Report - EAB-21:005137 Stockholm, Sweden June 2021.* Ericsson.

Ericsson. (2021). *Mobile Data Traffic Outlook.* Retrieved October 19, 2021, from https://www.ericsson.com/en/reports-and-papers/mobility-report/dataforecasts/mobile-traffic-forecast

ETSI. (2021). *5G security architecture and procedures for 5G System (3GPP TS 33.501 version 16.6.0 Release 16)–Technical Specification TS 133501 V16.6.0 (2021-04).* Retrieved from 5G Americas: https://www.5gamericas.org/wp-content/uploads/2021/01/InDesign-3GPP-Rel-16-17-2021.pdf

Farooq, M., Riaz, S., Abid, A., Umer, T., & Zikria, Y. (2020). Role of IoT technology in agriculture: a systematic literature review. *Electronics, 9*(2), 319.

Fontenla, E. (2016). Cooperativas agropecuarias argentinas: nueva generación de cooperativas. *Serie Documentos Nº 1; Ediciones CGCyM. Buenos Aires.*

Fonyi, S. (2020). Overview of 5G security and vulnerabilities. *International Conference on Cyber Conflict (CyCon US 2019) Defending Forward (Spring), 1*, pp. 117–134.

Forge, S., & Vu, K. (2020). Forming a 5G strategy for developing countries: a note for policy Makers. *Telecommunication Policy, 44*(7).

Frei, F. X., & Morriss, A. (2020). *Begin With Trust*. Retrieved October 10, 2021, from https://hbr.org/2020/05/begin-with-trust

Gabriel, C., & Kompany, R. (2021). *"Open RAN: ready for prime time? The operators' perspective, Research report, April 2021*. London, UK: Analysys Mason Ltd.

Gamreklidze, E. (2014). Cyber security in developing countries, a digital divide issue. *The Journal of International Communication, 20*(2).

GCSCC. (2021). *Cybersecurity Capacity Maturity Model for Nations (CMM) - 2021 Edition*. Global Cyber Security Capacity Centre Dpt. of Computer Science, University of Oxford, UK.

Gleißner, M., Dotzler, J., Hartig, J., Aßmuth, A., Bulitta, C., & Hamm, S. (2021). IT security of cloud services and IoT devices in healthcare. *Proc. XII International Conference on Cloud Computing, GRIDs, and Virtualization (Cloud Computing 2021)*.

Goedde, L., & Revellat, J. (2020). Agriculture's connected future: how technology can yield new growth. *Discussion Paper - McKinsey Global Institute, 2020*.

Goldman, J. G. (2014). *How being watched changes you – without you knowing*. Retrieved October 10, 2021, from https://www.bbc.com/future/article/20140209-being-watched-why-thats-good

Goussal, D. (2017). Rural Broadband in Developing Regions: Alternative Research Agendas for the 5G Era. In K. E. Skouby, I. Williams, & A. Gyamfi, *Handbook on ICT in Developing Countries: 5G Perspective*. Aalborg: River Publishers.

Grand View Research. (2021). *5G Infrastructure Market Size Share & Trends Analysis Report 2021–2028, Report ID: GVR-4-68038-234-1*.

Griffith, M. (2021). Balancing the Promise and the Peril of 5G: The State of Play of the United States. *5G Beyond Borders Workshop, Woodrow Wilson International Center*. USA.

GSMA Intelligence. (2019). *5G In SUb-Saharan Africa: laying the foundations*. Retrieved October 19, 2021, from https://data.gsmaintelligence.com/api-web/v2/research-file-download?id=45121572&file=2796-160719-5G-Africa.pdf

GSMA Intelligence. (2021, January). *The economics of mmWave 5G - GSMA Intelligence*. Retrieved October 12, 2021, from https://data.gsmaintelligence.com/api-web/v2/research-file-download?id=59768858&-file=210121-Economics-of-mmWave.pdf

Gupta, M., Abdelsalam, M., Khorsandroo, S., & Mittal, S. (2020). Security and privacy in smart farming: challenges and opportunities. *IEEE Access, 8* (2020).

HM Government. (2020, August 18). *5G mobile technology: a guide.* Retrieved October 12, 2021, from https://assets.publishing.service. gov.uk/government/uploads/system/uploads/attachment_data/ file/913179/5G_mobile_technology_a_guide.pdf

Hong, E., Ryu, J., & Lee, E. (2021). *Entering the 5g era: lessons from Korea.* Digital Development Global Practice, World Bank.

Horwitz, L. (2021). *How Industrial Edge Fuels Real-Time IoT Processes.* Retrieved October 19, 2021, from https://www.iotworldtoday. com/2021/03/18/how-industrial-edge-fuels-real-time-iot-processes/

Hurel, L. (2021). Cybersecurity in Brazil: an analysis of the national strategy. *Strategic Paper SP-54.*

IDB. (2020). Cybersecurity risks, progress, and the way forward in Latin America and the Caribbean. *Cybersecurity Report.*

IMDRF. (2020). Principles and Practices for Medical Device Cybersecurity. *International Medical Device Regulators Forum, Doc. IMDRF/CYBER WG/N60FINAL:2020.*

Immerman, G. (2021). *The Importance of Edge Computing for the IoT.* Retrieved October 10, 2021, from https://www.machinemetrics.com/ blog/edge-computing-iot

ISO/IEC. (2012). *Information security techniques- Guidelines for cyber-security- ISO/IEC 27032.* Retrieved from International Organization for Standardization (ISO)/International Electrotechnical Commission (IEC): https://www.iso.org/standard/44375.htm

ITU. (2020). *Global Cybersecurity Index 2020 - Measuring commitment to cybersecurity.* Geneve: ITU-D (International Telecommunications Union).

Joy Online. (2021). *Ghana Police Service has 18,000 convicts on database – IGP.* Retrieved October 19, 2021, from https://www.myjoyonline.com/ ghana-police-service-has-18000-convicts-on-database-igp/?param=

Khan, R., Kumar, P., Jayakody, D., Dushantha, N., & Liyanage, M. (2019). A survey on security and privacy of 5G technologies: potential solutions, recent advancements and future directions. *IEEE Communications Surveys & Tutorials, 22*(1), 196–248.

Koebler, J. (2017). *Why American Farmers Are Hacking Their Tractors With Ukrainian Firmware.* Vice Mother board.

Køien, G. M. (n.d.). On threats to the 5G service based architecture. *Wireless Personnal Communications, 119*, 97–116.

Kristen, E., Kloibhofer, R., Díaz, V., & Castillejo, P. (2021). Security Assessment of Agriculture IoT (AIoT) Applications. *Appl. Sci., 11*(13), 5841.

Kumar, P., & Sharma, S. (2021). An empirical evaluation of various digital signature schemes in wireless sensor network. *IETE Technical Review, 2021.*

Lieder, S., & Schröter-Schlaack, C. (2021). Smart farming technologies in arable farming: Towards a Holistic Assessment of Opportunities and Risks. *Sustainability, 13*(2021), 6783.

Lumbard, K., Ahuja, V., & Snell, M. (2020). Open Agriculture and the Right-to-Repair Community Movement. *MWAIS 2020 Proc.15th Midwest Association for Information Systems Conference, Des Moines, Iowa May 28–29.*

Maquinac. (2021). *La pulverización selectiva se afianza en Argentina.* NVS Desarrollos.

Matinmikko-Blue, M., Yrjölä, S., Ahokangas, P., & Hämmäinen, H. (2021). Analysis of 5G spectrum awarding decisions: How do different countries consider emerging local 5G networks? *23rd Biennial Conference, International Telecommunications Society (ITS) Sweden 21st–23rd June, 2021.* Gothenburg.

MDCG. (2019). Guidance on cybersecurity for medical devices. *Medical Device Coordination Group - Doc. MDCG 2019-16, (EU) 2017/745, 2019.*

Meyerhoff, R. (2019). *Argentina gets smarter about sustainable agriculture.* Forbes.

Monzon, L. (2021). *MTN Announces 5G Roll Out Date in Ghana.* Retrieved October 19, 2021, from https://www.itnewsafrica.com/2021/09/mtn-announces-5g-roll-out-date-in-ghana/

Mueller, K., Coburn, A., Knudson, A., Lundblad, J., McBride, T., & MacKinney, C. (2021). *Characteristics and challenges of rural ambulance agencies–A brief review and policy considerations.* USA: Rural Policy Research Institute (RUPRI).

Nai-Fovino, I., Neisse, R., Hernandez-Ramos, J. L., Polemi, N., PoRuzzante, G., Figwer, M., et al. (2019). A Proposal for a European Cybersecurity Taxonomy. *EUR 29868.*

National Communication Authority. (2021, June 1). *INDUSTRY INFORMATION - TELECOM SUBSCRIPTIONS FOR MARCH 2021.* Retrieved October 19, 2021, from https://www.nca.org.gh/assets/Industry-Report-March-2021-.pdf

National Communications Authority & Ghana Statistical Service. (2020, March). *Household Survey on ICT in Ghana.* Retrieved October 12,

2021, from https://statsghana.gov.gh/gssmain/fileUpload/pressre-lease/Household%20Survey%20on%20ICT%20in%20Ghana%20(Abridged)%20new%20(1).pdf

NIST. (2018). *Framework for improving critical infrastructure cybersecurity, Version 1.1.* USA: National Institute of Standards and Technology.

Nowak, T., Sepczuk, M., Kotulski, Z., Niewolski, W. A., Artych, R., Bocianiak, K., et al. (2020). Verticals in 5G MEC-Use Cases and Security Challenges. *IEEE Access, 9*(2020).

NUKIB. (2018). The Prague proposals-chairman statement on cybersecurity of communication networks in a globally digitalized world. *European Union 5G Security Conference, Prague 3 May 2019- National Cyber and Information Security Agency (NÚKIB).* Czech R.

Racovita, M. (2021). *Industry briefing: cybersecurity for the Internet of Things and Artificial Intelligence in the AgriTech sector.* London, UK: Industry Briefing PETRAS National Centre of Excellence for IoT Systems Cybersecurity.

Roddy, M., Truong, T., Walsh, P., Bado, M., Wu, Y., Healy, M., et al. (2019). 5G Network Slicing for Mission-critical use cases. *2019 IEEE 2nd 5G World Forum (5GWF),* (pp. 409–414).

Rose, D., Wheeler, C., Winter, M., Lobley, M., & Chivers, C. (2021). Agriculture 4.0: Making it work for people, production, and the planet. *Land Use Policy, 100.*

Rugeles, J., Guillen, E., & Cardoso, L. (2021). A technical review of wireless security for the Internet of Things: Software Defined Radio perspective. *Journal of King Saud University - Computer and Information Sciences.*

Russom, P. (2018). *Data Requirements for Machine Learning.* Retrieved October 10, 2021, from https://tdwi.org/articles/2018/09/14/adv-all-data-requirements-for-machine-learning.aspx

Samsung. (2019). 5G Launches in Korea: get a taste of the future. *White Paper.*

Scaramuzza, F. (2021). Adelantos tecnológicos y el futuro inmediato de la agricultura de precisión. *(Conference) INTA (Instituto Nacional de Tecnología Agropecuaria. Manfredi, Córdoba. Biblioteca Fundación CIDETER, Argentina 2021.*

Shimabukuro, I. (2021, 12 5). *Brazilian government inaugurates its first 5G antenna in a rural area of the country.* Retrieved from Olhar Digital: Brazil.

Świątkowska, J. (2020). Tackling cybercrime to unleash developing countries' digital potential pathways for prosperity. *Commission Background Paper Series, 33.*

Talukder, A., & Hass, R. (2021). AIoT: AI meets IoT and web in smart healthcare. *WebSci'21 Companion, June 21–25, 2021, Virtual Event, UK.*

Thakor, V., Razzaque, M., & Khandaker, M. (2021). Lightweight cryptography algorithms for resource-constrained IoT devices: A review, comparison and research opportunities. *IEEE Access, 9*(2021).

Thales. (2021). *5G technology and networks (speed, use cases, rollout).* Retrieved October 10, 2021, from https://www.thalesgroup.com/en/markets/digital-identity-and-security/mobile/inspired/5G

Tomasin, S., Centenaro, M., Seco-Granados, G., Roth, S., & Sezgin, A. (2021). Location-privacy leakage and integrated solutions for 5G cellular networks and beyond. *Sensors, 21*(2021), 5176.

Trakadas, P., Sarakis, L., Giannopoulos, A., Spantideas, S., Capsalis, N., Gkonis, P., et al. (2021). A cost-efficient 5G non-public network architectural approach: key concepts and enablers, building blocks and potential use cases. *Sensors, 21*(2021), 5578.

Tursunov, R., Lenox, J., & Cleave, J. (2019). *Digital healthcare South Korea. Market intelligence report 2019.* UK: Department for International Trade (DIT) - Intralink Group.

Unit 42. (2020). *2020 Unit 42 IoT Threat Report.* CA USA: Palo Alto Networks, Santa Clara.

USTDA. (2020). *ICT project opportunities in Argentina, Brazil and Paraguay. A resource guide for US industry.* USA: U.S. Trade and Development Agency.

Wazid, M., Kumar, A., Shetty, S., Gope, S., & Rodrigues, J. (2021). Security in 5G-enabled Internet of Things communication: issues, challenges and future research roadmap. *IEEE Access, 9*(2021).

Weissman, C. (1969). Security control in the ADEPT-50 time-sharing system. *Proc. 35th AFIPS Conference*, (pp. 119–133). USA.

Winkler, K., Fuchs, R., Rounsevell, M., & Herold, M. (2021). Global land use changes are four times greater than previously estimated. *Nature Communications, 12*(2501).

WTI. (2020). *Annual Report 2020.* World Tele Health Initiative, Santa Barbara, CA. USA.

Wu, T., Yang, L., Lee, Z., Chu, S., & Kumar, S. (2021). A provably secure three-factor authentication protocol for Wireless Sensor Networks. *Wireless Communications and Mobile Computing.*

Yazdinejad, A., Zolfaghari, B., Azmoodeh, A., Dehghantanha, A., Karimipour, H., Green, A. G., et al. (2021). A review on security of smart farming and precision agriculture: security aspects, attacks, threats and countermeasures. *Applied Sciences, 11*(16), 7518.

Zhai, Y., Xu, X., Chen, B., Lu, H., Wang, Y., & Li, S. S. (2020). 5G Network-enabled smart ambulance: architecture, application, and evaluation. *IEEE Network*.

Zhang, A., Heath, R., McRobert, K., & Llewellyn, R. (2021). Who will benefit from big data? Farmers' perspective on willingness to share farm data. *Journal of Rural Studies*.

1

Expansion–Security Tradeoffs in the Pathway to Rural 5G Networks

Darío M. Goussal

School of Engineering, Northeast University at Resistencia (UNNE)-
Resistencia, Argentina
Email: dgoussal@yahoo.com

Abstract

In Latin America and the developing world, 5G networks are already deployed in some urban areas, although the critical decisions on spectrum allocation and security are still pending. In 2019, the 5G Security Conference released a bundle of recommendations "The Prague Proposals" that shook the entire mobile technology industry, back then immersed in the China–USA commercial war. Yet the next generation of rural networks will face security and privacy challenges, far from the current priorities of regulators and giant tech providers. The main threat will likely have another source: an uneven expansion of new applications like e-farming and remote healthcare, with myriads of devices other than normal handsets requiring stringent reliability standards and resilient capabilities. Developing nations will need customized introduction strategies seeking suitable tradeoffs between two extremes: early rural coverage to rapidly foster the economy at the risk of untrusted solutions and unproven business cases, or maximum security via delayed expansion and cautious investment plans, at the risk of losing the IoT leapfrogging opportunity. The pathway to 5G in developing countries will show a two-speed rollout landscape, where multi-access edge computing (MEC) verticals like smart farming/agribusiness and digital healthcare may overtake national ecosystems and conventional mobile network operator (MNO) expansion strategies. Despite urban-centred 5G forecasts, rural niches will not need to wait that long: it will ultimately depend on the ability of regulators and concerned stakeholders in solving the puzzle of new alliances, private and stand-alone

19

strategies, spectrum, and resilience policies to unleash suitable applications ahead of the expected rise of cybersecurity threats.

1.1 Introduction

In Latin America and the developing world, the strength and productivity of agribusiness, international food value chains, and other tech-dependent sectors are rooted in the countryside. People living in small communities, scattered spots, and farms need to work, study, and move therein. Efficient and timely access to commerce, government services, healthcare, and education is increasingly requiring reliable, state-of-the-art telecommunications networks deployed across large geographical areas. Mobile networks must be secure and private because of the value of information assets they are entrusted to transfer. Cybersecurity as a discipline evolved since the 1960s when a number of security-penetration analyses performed in projects of the U.S. Department of Defense (DoD) with system providers, gave rise to formal studies on the evaluation of software integrity and audit approaches (Weissman, 1969).

However, in telecommunications systems, the assessment and response to digital security or privacy violations is also a matter of social sciences and law enforcement demands. The changing dynamics of cybercrime and its worldwide spreading reflects also changes in society, particularly the increasing use of broadband-appetite applications. Since 2020, the forced confinement in many countries due to COVID-19 pandemics expanded the adoption and time use of video conferencing, virtual education, remote healthcare, home office/teleworking and entertainment, all heavy-data rate, and low-latency services.

Along the years, rural telecommunications and cybersecurity have shared a common peculiarity: their holistic, interdisciplinary, and inter-temporal nature. In other publications about the 5G era, we have referred to the long-lasting need for telecommunications in rural developing regions - currently in their third century of expansion efforts- and the blend of technical, economic, regulatory, and socio-cultural aspects shown in their research agendas (Goussal, 2017). According to the High Level Advisory Group of the European Commission (EC),*"cybersecurity is not a clearly demarcated field of academic study that lends itself readily to scientific investigation"*. Rather, cybersecurity combines a multiplicity of disciplines from the technical to behavioral and cultural. Scientific study is further complicated by the rapidly evolving nature of threats, the difficulty to undertake controlled experiments, and the pace of technological change and innovation. In short, cybersecurity is much more than a science" (Nai-Fovino et al, 2019)

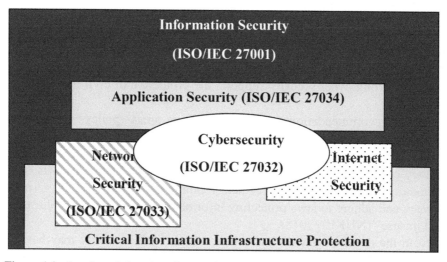

Figure 1.1 Interleaved domains of network, application, and Internet security (**ISO/IEC, 2012**).

In the same vein, another report from a giant multinational provider state: "Building a secure 5G requires us to take a holistic view and not only focus on individual technical parts in isolation. For example, interactions between user authentication, traffic encryption, mobility, overload situations, and network resilience aspects need to be considered together. It is also important to understand relevant risks and how to address them appropriately" (Ericsson, 2018).

Information security has been defined as "preservation of confidentiality, integrity, and availability of information" (ISO/IEC 27001), while cybersecurity is the protection of such attributes for information assets in the cyberspace, as reflected in the interleaved domains of network, application, and Internet security (ISO/IEC 27032, Figure 1.1) (ISO/IEC, 2012).

In 2007–2008, successive cyberattacks in Estonia and Georgia featuring warfare characteristics ended the era of "tech-optimism" in global telecommunications networks and raised the digital insecurity threat to an upper layer. Some authors argued that the contingent status of cybersecurity in developing countries, their resilience and recovery capabilities is another expression of the digital divide and reflects the sound differences in ITC between nations (Gamreklidze, 2014; Świątkowska, 2020; Griffith, 2021).

In an attempt to raise cybersecurity from its customary technical level to an upper, more holistic category and to build consensus among 5G stakeholders, a document released after the EU Security Conference in 2019

("The Prague Proposals") highlighted: "Cyber security cannot be regarded as a purely technical issue. A safe, secure, and resilient infrastructure requires adequate national strategies, sound policies, a comprehensive legal framework, and dedicated personnel, who is trained and educated appropriately. Strong cyber security supports the protection of civil liberties and privacy".

The recommendations encompassed four areas (policy, technology, economy and security/privacy/resilience) with tacit references to the China–USA dispute about trusted 5G supplies, even when the text recognized: "Every country is free in accordance with international law, to set its own national security and law enforcement requirements, which should respect privacy and adhere to laws protecting information from improper collection and misuse" (NUKIB, 2018).

In the developing world, a holistic vision of 5G security may have a different background and different priorities. In short, there is a lot of pending work in the implementation of the Prague agenda to the legacy mobile ecosystem, its current landscape, applications, and regulations prior to undertaking their best strategies for 5G security. In this chapter, we will examine technical and policy aspects of 5G cybersecurity in rural environments on two prospective use cases, considered as concrete needs but exposed to major cyber risks in Latin America: e-healthcare and smart agriculture. We argue that each expansion strategy targeting such needs will comport a customized tradeoff between two extremes: an early rollout in absence of trusted solutions and proven business models to rapidly boost productivity and access to e-services, or a delayed plan with cautious investments and guaranteed supplies, at the risk of losing the IoT leapfrogging opportunity (GCSCC, 2021).

A detailed analysis of potential 5G technical vulnerabilities is out of the scope of this article. Our intention is to discuss some risks and tradeoffs associated with the prospective expansion of 5G networks in rural areas throughout sample use cases and relevant facts. The rest of this chapter is organized as follows: Section 1.2 presents a little-known fact: the paradox of early rural 5G deployments in developing countries. Section 1.3 describes international cybersecurity frameworks and benchmarking indexes. Section 1.4 is a briefing on technical aspects of 5G referred to security and privacy aspects. Sections 1.5 and 1.6 discuss opportunities and risks in two envisaged verticals, respectively smart farming and rural healthcare, along with their current status, expansion needs, and national plans. Section 1.7 presents a vision of expansion-security tradeoffs in developing countries in hypothetical scenarios. Section 1.8 refers to user privacy and other overlooked issues, and Section 1.9 explains the conclusions.

1.2 Rural Networks and the Expansion Paradox

The conventional wisdom about 5G expansion is that of a purely urban business in densely populated cities with incumbent providers seeking profitability on the basis of quantity: millions of smartphones, tablets, and laptops using heavy-data rate virtual meeting applications, GPS maps, real-time streaming platforms, or HQ video games, while daily commuting to schools and offices. The rationale of such envisioned model is that of traditional MNOs choosing a retrofit strategy ("brownfield deployment") on major cities' downtown areas: the upgrade of legacy 4G/LTE infrastructure through a 5G Non-Standalone Architecture (NSA). This has been the preferred approach since 2018, in the 61 countries with active 5G operators accounting for a revenue share of 92.9% in 2020 (Grand View Research, 2021).

The strange fact in 2021 is an unforeseen expansion paradox: rural networks already deployed, pioneering the introduction of the new mobile generation, even in developing countries. Moreover, there are pilot trials of 5G-enabled, advanced IoT applications in rural spots, out of the industrialized world. In Brazil, the first three 5G rural networks were activated in 2021 in Rondonópolis, MT (Mato-Grossense Cotton Institute); Sorocaba, SP (Facens University Center); and Londrina, PR. In the first case, a model farm with a cotton plantation is remotely monitored with high-resolution images taken by drones, with 5G equipment provided by Nokia. The Sorocaba campus was chosen to demonstrate a remote-activated irrigation system with sprayers connected to the network and commanded from a regular 5G handset (Moto G100), besides a pilot cattle remote veterinary monitoring, and other agritech IoT projects. The third rural trial was set up in August 2021 in Embrapa-Soja, a soy experimental farm owned by federal agriculture research corporation Embrapa in Londrina, in the state of Paraná. This facility was chosen to test a 3.5 GHz native Stand Alone 5G public network, installed by the local operator Sercomtel with a temporary, experimental spectrum license issued by Anatel, the federal regulatory body (Shimabukuro, 2021; Cordeiro, 2021).

The rural expansion paradox has three important drivers. First, the promise of 5G brought a mix of new players, new operator–vendors alliances, new IoT applications, and new deployment strategies. In Australia, Telstra's 5G network boasts to have launched 5G services in over 2700 suburbs and 200 towns or cities reaching rural areas and achieving a world record long-distance 5G link – a 113 km-span from a single mobile tower. Telstra has decided to use carrier aggregation technologies with low-band (850 MHz) and mid-band (3.6 GHz) combined spectrum, aimed at enhancing long-range coverage in suburban and rural areas (Lewis, 2021).

The second driving force is the lower risk and easiness of rollout of 5G in rural and enterprise environments, in particular for MNOs prone to virtualization and/or adopting Open RAN strategies (O-RAN). According to a survey of a hundred operators in 3Q 2020, O-RAN will make faster progress in rural and enterprise scenarios, where many multi-vendor networks are currently being tested (for instance, Vodafone is trailing networks in diverse rural settings, from Ireland to Congo). These markets are likely to feel the impact of O-RAN before others because the claimed objectives of reduced cost and supplier diversity look achievable at first in these niches as a result of relatively low data loads and limited coverage areas: only Rakuten (Japan) has deployed a large-scale O-RAN commercial network and Dish (USA) is just seeking the same. MNOs with a not-urgent timescale for virtualizing the RAN can trial in enterprise or rural scenarios while maintaining traditional architecture in their primary RAN Commercial buildout in greenfield, rural and enterprise markets are expected to begin in the near term (1–3 years), while urban macro deployments using 3.5 GHz or millimeter wave, MMIMO, and carrier aggregation are forecasted toward 2023 at the earliest. Table 1.1 depicts the perceived challenges for the adoption of O-RAN strategies, where "Security issues" and "Real-time resilience" are obviously among them. The "Rural and Enterprise" scenario ranks fairly well throughout these 10 challenges, with perceived risk/benefit ratios 1 (Low) for resilience and 2 (Neutral) for security (Gabriel & Kompany, 2021).

The third driver – and likely the most important – is a true need to start using rural 5G applications in some industry verticals as soon as possible, in order to leverage existing competitive advantages and to cope with costs and risks other than cybersecurity issues (even when end users ignored tech aspects). Smart farming and rural healthcare are regarded as feasible 5G verticals, even in developing countries.

Concerning the first one, Latin American countries such as Argentina and Brazil have a long-standing tradition as large agribusiness producers, exporters, and early adopters of precision agriculture technologies including exports in agricultural machinery industries and control devices. Research in precision agriculture by government agencies such as Argentina's National Agricultural Technology Institute (INTA) started around 1995 in digital crop mapping and satellite-guided applications for direct seeding. Now INTA delivers courses every year for 450 attendants in the Manfredi agricultural testing station (autopilot systems, GPS receivers, mapping, yield monitor systems, drones, NDVI imaging, weed sensors, automatic section cutting, variable rate seeding, Blockchain, Big data, and AI). Precision spraying

Table 1.1 The perceived challenges for the adoption of O-RAN strategies.

Challenge	Primary macro network	Rural or enterprise network	Greenfield macro net sub-3Ghz	Mission-critical network
Integration cost and complexity	3	2	2	3
Immature technologies/ supply chain	3	1	2	3
Performance tradeoffs over integrated RAN	3	2	2	3
Risk of lowest common denominator	2	1	2	3
Lack of scale in the ecosystem	2	1	1	2
Real-time resilience	3	1	3	3
Security issues	3	2	3	3
Risk of increased OPEX	3	2	2	1
Lack of single accountable supplier	2	1	2	2
Uncoordinated update processes lifecycle management challenges	1	1	1	2

Notes:
1. Rewards outweigh risks; risks are non-critical to the business case
2. Neutral – there are significant rewards and operators feel able to mitigate the risks
3. Risks outweigh rewards until they are effectively mitigated.

weed sensors manufactured in Argentina in last years experienced annual sales growth averaging 20%, with reported savings of up to 90% in inputs. International sales of agricultural precision machinery to the United States, Australia, and the EU expanded with self-propelled smart sprayers exported to South America, Africa, and Central Asia (Maquinac, 2021).

Argentina and USA have a high degree of geographical spreading of farms in small and medium production units, traditionally grouped in agro-co-operatives. Since 2019, the Argentine Cooperatives Association (ACA) with around 150 associated cooperatives and 50,000 farmers has a digital farm platform that combines geospatial data from satellites and drones to monitor ambient soil with weather and business data in order to forecast the potential productivity of sown lots. The platform runs machine learning algorithms using such data to generate analytic information and specific recommendations for the associated farmers. ACA, which actually accounts for some 17%

of the national crop exports, expects productivity gains of some 10% by using this platform (Meyerhoff, 2019; Fontenla, 2016).

1.3 International Benchmarks: Cybersecurity Indexes and Frameworks

In the last half of the 2010 decade, the spreading of large-scale cyberattacks at national and international levels has stressed the need for moving up cybersecurity and privacy to far-reaching places. The challenge is building up stronger consensus on regulatory and policy agendas by translating law enforcement mandates into tangible, working instruments. Legislation on digital security and privacy is already enforced in several nations, along with specific sector regulations e.g., in the banking industry. But citizens in previous years realize that laws and institutional rulemaking geared just to apply penalties will not deter cybercrime. In cybersecurity strategies "one size does not fit all", as seen in the unevenness of budgets, expert workforces, and the prevailing level of security readiness in the developing world. For example, in February 2020, Brazil approved its first Cybersecurity Strategy (E-Ciber), establishing the main actions to be carried out by the government in 2020–2023. But in November of the same year, there was a major attack on the Superior Court of Brazil (TSJ), labeled as "the worst cyberattack in the country's history". The ransomware encrypted all the files in the judicial system's court, once more demonstrating the systems' failures and the Federal Public Administration's lack of preparedness in responding to these attacks (Hurel, 2021).

A concrete strategy for securing any network, website, ITC device or application at any scale, from home offices to enterprises and government institutions, should provide means for determining digital weaknesses, degrees of risk, appropriate defenses, and recovery procedures. The need to rank and evaluate progress in common dimensions and indicators has brought dynamic security measures, security rankings, and cybersecurity frameworks as partial responses to such demands. Now there are near 30 cybersecurity frameworks of different types and levels, including national and international standards, indexes, and models proposed by non-profit institutions, government expert agencies, academic or mixed corporations. They share common characteristics: multi-dimensional structure, dynamic self-evaluation processes, and progressive stages of attainment. For the sake of brevity, we will refer just to three of them (CMM, GCI, and NIST).

The Cybersecurity Capacity Maturity Model for Nations (CMM), developed in the University of Oxford through its Global Cyber Security

Capacity Centre (GCSCC) in consultation with more than 200 international experts drawn from governments, academia, and society, is an assessment tool focusing on cumulative goals. It provides monitoring of maturity levels in national cybersecurity capabilities, along five dimensions: (1) developing cybersecurity policy and strategy, (2) encouraging responsible cybersecurity culture within society, (3) building cybersecurity knowledge and capabilities, (4) creating effective legal and regulatory frameworks, and (5) controlling risks through standards and technologies. The degree of cybersecurity attainment is ranged in five stages, reflecting successive maturity status in a sort of roadmap (GCSCC, 2021; IDB, 2020).

- Start-up: No cybersecurity maturity exists, or it is very embryonic in nature. Beyond the first discussions on cybersecurity capacity building, no concrete actions have been taken. No observable evidence of cybersecurity capacity.

- Formative: Some aspects have been formulated and initiated with evidence clearly demonstrated, but they are rather ad-hoc, disorganized, poorly defined, or simply new.

- Established: The elements of a dimension are in place, defined, and functioning. However, there is no well thought-out consideration of the relative allocation of resources. Little tradeoff decision-making concerning the investment involved.

- Strategic: Choices have been made about which indicators are more and less important for the organization or state under particular circumstances.

- Dynamic: Clear mechanisms in place to alter strategy depending on the prevailing circumstances e.g., technological sophistication of the threat environment, global conflict, or significant change in one area of concern (e.g., cybercrime or privacy). Dynamic organizations have developed methods for changing strategies in stride. Rapid decision making, reallocation of resources, and constant attention to the changing environment are features of this stage.

The CMM database was the source for an empirical study on the social and cultural aspects of cybersecurity capacity building in 78 nations. The sample included 47 countries reviewed once by the GCSCC during the period 2015–2020, and 31 countries surveyed by GCSCC in collaboration with the Organization of American States (OAS) and the Inter-American Development Bank (IDB), using an online survey completed by OAS member states. The linear regression estimates the social and cultural dimension (SCD) of cybersecurity maturity for countries by two explanatory variables obtained by factor analysis that were found as significant

$$y_i = 1.73 + 0.34 \, Development + 0.18 \, Scale, \tag{3.1}$$

where "Development" and "Scale" (combinations of seven secondary variables of development and scale of Internet use) were determined as the closest explanatory variables of cultural attitudes, values, and practices of societies tied to cybersecurity. The model enables comparison between actual and predicted values – which countries in the CMM database show significantly higher or lower maturity than predicted. Regional effects were not found as statistically significant (Creese, Dutton, & Esteve-González, 2021).

In 2014 in the United States, the Cybersecurity Enhancement Act (CEA) expanded the role of the National Institute of Standards and Technology (NIST) toward developing cybersecurity risk frameworks for voluntary use by critical infrastructure owners and operators. NIST must identify "a prioritized, flexible, repeatable, performance-based, and cost-effective approach, including information security measures and controls that may be voluntarily adopted by owners and operators of critical infrastructure to help them identify, assess, and manage cyber risks" (CEA, 2014).

Later in 2018, NIST released version 1.1 of the Framework, which adds significant content in authentication, identity, security management, and self-assessment, focusing on the organization's risk handling processes. It provides a common taxonomy and self-evaluation mechanisms for organizations to:

1.   Describe their current cybersecurity posture

2.   Describe their target state for cybersecurity

3.   Identify and prioritize opportunities for improvement within the context of a continuous and repeatable process.

4.   Assess progress toward the target state

5.   Communicate among internal and external stakeholders about cybersecurity risk.

The framework has three parts: Core, implementation tiers, and profiles. The core is a set of common cybersecurity activities, outcomes, and informative references across sectors and critical infrastructure. Detailed guidance for developing organizational profiles arise from elements of the core, while the use of profiles helps an organization to align and prioritize its cybersecurity activities with its own resources, business or mission requirements, and risk tolerances. The tiers provide a mechanism to view and understand the characteristics of each organization to manage risks and help it in prioritizing and achieving security objectives. However, since the identification of cybersecurity performance metrics is evolving, the Framework has a caveat: "organizations should be thoughtful, creative, and careful about the ways in which they employ measurements to optimize use, while avoiding reliance on artificial indicators of current state and progress in improving cybersecurity risk management. Judging cyber risk requires discipline and should be revisited periodically" (NIST, 2018).

The International Telecommunications Union (ITU) launched in 2007 its Global Cybersecurity Agenda (GCA), wherein the Global Cybersecurity Index (GCI) is an evaluation framework that measures the degree of engagement of nations to pursue a cybersecurity agenda. It has 20 indicators constructed through 82 questions of an on-line survey and grouped around five dimensions or "pillars" (Legal, Technical, Organizational, Capacity Building, and Cooperative Measures). As a composite weighted index, each indicator, sub-indicator, and micro-indicator is assigned a weight, reflecting its relative importance. Weights have a significant impact on final scores, and different techniques will produce different rankings: the GCI took a participatory approach, using the budget allocation process (BAP). This fundamentally considers weights as value judgments, so a wide variety of expert input is a must. In the 2020 edition, 84 experts were asked to provide weightage recommendations in pillars related to their respective areas of expertise.

Legal measures (including legislation, regulation, and containment of spam legislation) authorize a state to set up basic response mechanisms through investigation and prosecution of crimes, and the imposition of sanctions for non-compliance or breach of law. The Technical pillar reflects the existence of technical institutions and framework dealing with cybersecurity. Countries require accepted minimum-security criteria and accreditation schemes for software applications and systems, complemented with a national body to manage cyber incidents, an authoritative government entity and a national framework to watch, warn, and respond to incidents. The Organizational score encompasses the existence of coordinating institutions, policies, and strategies for cybersecurity development at the national level. Organizational measures include the identification of cybersecurity objectives and strategic plans, as well as the formal definition of institutional roles, responsibilities, and accountabilities to ensure their implementation. The Capacity pillar assesses research and development, education and training programs, certified professionals and public sector agencies fostering capacity building. Finally, Cooperative measures assess the existence of partnerships, cooperative frameworks, and information sharing. The unprecedented level of interconnection between states makes cybersecurity a shared responsibility and a transnational challenge. The latest GCI report (2020) shows sound differences in score values between developed and developing countries, regardless of their global region, as depicted in Table 1.2 for the Americas' best-ranked states (ITU, 2020).

1.4 Technical Foundations

In 1972, the U.S. Air Force commissioned a study on the protection of classified military/government information in computer systems ("The Anderson Report"), which identified three categories of potential security risks: (1) unauthorized information release, (2) unauthorized information modification, and (3) unauthorized denial of use. That was the root of the so-called CIA triad (Confidentiality, Integrity, and Availability), mentioned in the Preamble section of the Prague Proposals: "Security of communication channels has therefore become vital. Disruption of the integrity, confidentiality or availability of transmitted information or even the disruption of the service itself can seriously hamper everyday life, societal functions, economy and national security" (NUKIB, 2018).

There are threats with different hazard levels and varying impacts on all these dimensions. The literature has many studies and reviews on the CIA model applied to 5G, and since most attacks are old types – they came from the 4G era –, a detailed behavioral analysis is useless. Rather we will focus on the security architecture of 5G and the potential risks relevant to rural 5G use cases. The fundamental change introduced in the shape of the 5G core network is the Services Based Architecture (SBA), a topology consisting of a common bus with three planes (layers), each of them conveying different types of traffic: (a) the user plane for data, (b) the control plane for signaling, and (c) the management plane for administrative communications (Figure 1.2) (Ericsson, 2018).

This common bus interconnects functional elements (NEF, network exposure; NRF, network repository; NSSF, network slice selection; PCF, policy control; AMF, access and mobility; AUSF, authentication server, SMF, session management; SCP, service communication proxy; AF, application). Direct and indirect communication and delegated discovery are enabled through the Service Communication Proxy. The control plane functionality and common data repositories are delivered by way of a set of interconnected

Table 1.2 GCI report (2020).

Countries	Overall	Dimensions				
		Legal	Technical	Organization	Capacity development	Cooperative measures
USA	100.00	20.0	20.0	20.0	20.0	20.0
Canada	97.67	20.0	18.3	20.0	20.0	19.4
Brazil	96.60	20.0	18.7	19.0	19.5	19.4
Mexico	81.68	15.6	17.9	14.7	16.1	17.3
Uruguay	75.15	13.9	18.3	12.1	19.0	11.8
Dominican Rep.	75.07	16.4	18.4	18.5	9.9	11.8
Chile	68.83	17.2	9.4	15.8	11.1	15.3
Costa Rica	67.45	17.6	9.1	12.7	12.1	15.9
Colombia	63.72	9.1	17.6	6.7	14.4	15.9
Cuba	58.76	14.9	10.9	13.9	10.5	8.6
Paraguay	57.09	14.2	10.9	13.1	6.8	12.1
Peru	55.67	20.0	11.6	5.6	5.3	13.2
Argentina	50.12	12.2	13.8	8.3	4.4	11.6
Panama	34.11	10.4	10.9	2.4	6.1	4.3
Jamaica	32.53	11.5	2.2	7.9	6.7	4.3
Suriname	31.20	11.1	7.0	1.7	7.1	4.3
Guyana	28.11	13.1	2.5	6.5	2.2	3.8
Venezuela	27.06	8.8	7.7	6.2	4.4	0.0
Ecuador	26.30	10.2	9.6	0.0	6.5	0.0

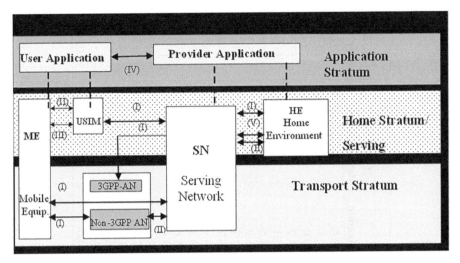

Figure 1.2   5G-overview of the security architecture.

network functions, each with authorization to access each other's services or sets of services. The core provides the mechanism to convert today's IP multi-media subsystem (IMS) to use of a service-based architecture (SBA) that provides flexibility and scale in service delivery as well as support for new capabilities such as network slicing (NS) (Roddy et al., 2019). NS is a new capability of 5G core to give flexibility to network services and applications to a traffic of different type, characteristics, and service-level agreements. Virtualized resources to manage specific groups of subscribers include a dedicated User Plane Function (UPF), Session Management Function (SMF), and Policy Control Function (PCF). When referring to cybersecurity threats in networking ecosystems, there are boundaries sometimes fuzzy between the telecommunications network and its applications. Historically, security standards in mobile networks have been under the umbrella of 3GPP specifications, whose first set for 5G (Releases 15–16) focused on the lower layers (NR, New Radio).

In 2018, the 3rd Generation Partnership Project began the Work on NR Release 16, finalized in July 2020. The features introduced as part of NR Release 16 target new verticals (transport industry, manufacturing, enterprise, industrial IoT, automobile, etc.) as well as enhanced capacity and operational efficiency (ETSI, 2021; CSRIC 7, 2020).

Figure 1.3 depicts the 5G security domains. Besides legacy features brought from 4G/LTE and previous generations (confidentiality of end user identifiers, authentication, encryption, etc.), there are three new elements:

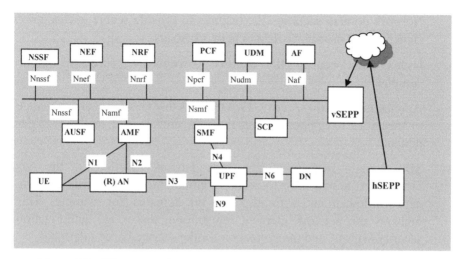

Figure 1.3 5G system architecture with Security Edge Protection Proxy (SEPP).

- International Mobile Subscriber Identity (IMSI): information encryption function to protect subscriber information. User privacy was improved by IMSI encryption, a key embedded in the code inside the SIM card, which encrypts the long-term identifier code before its wireless transmission. Thus, the contingent location and number of each subscriber are integrity protected and subject to mutual authentication, making difficult the "who and where" traceability without the decryption key

- Security Edge Protection Proxy (SEPP): intended to solve Signaling System 7 security issues between roaming domains and to implement the application layer security function between different carriers (i.e., public land mobile network)

- Integrated authentication: can use the same authentication method for 3GPP and non-3GPP accesses. SEAF enables devices to be re-authenticated without executing a full authentication method (e.g., AKA), even when they move between different access networks or between different service networks (CSRIC 7, 2020).

Security domains are groups of elements supporting each a set of related functions, namely:

- Network access security (I): enables secure authentication of access and network services, either from 3GPP or non-3GPP access networks to user's mobile equipment.

- Network domain security (II): features to provide secure exchange of signaling and user plane data to network nodes.

- User domain security (III): features to secure user access to mobile equipment.

- Application domain security (IV): features that enable secure exchange of messages to applications in the user domain and in the provider domain.

- SBA domain security (V): enables network functions in the services-based architecture to securely communicate within the serving network domain and with other network domains (for roaming). This includes the network element registration, discovery, authorization security aspects, and the protection for the service-based interfaces.

Frequent privacy threats like eavesdropping with rogue cell towers and IMSI catchers have been reported in different incidents in 4G/LTE and other legacy systems that use the Public Key Infrastructure (PKI). Such authentication procedure uses a cryptographic key embodied in each SIM card, which allows access to the network just after successful handshaking of digital certificates between the base station and the user. In 5G, this approach can no longer continue due to a combined security-privacy threat: in the case of a large-scale incident e.g., a massive theft of IMSI codes, it would be impossible to replace SIM cards of millions of devices (as will occur in Use Equipement's - UEs- in the case of IoT). Besides that, since the stealth detection of an IMSI code by rogue cell towers/ IMSI catchers entails the possibility of tracking the instant location, displacements, and private habits of the user, considerable efforts have been made to set up a new encrypted, worldwide common PKI procedure (Fonyi, 2020).

Nevertheless, end-to-end security and privacy are still open issues because they were not addressed in the standards from the very beginning. Historically, 3GPP specifications aimed at securing interoperability among mobile networks with a focus on connectivity functions, and did not attempt to cover security or privacy in Internet applications – except for Voice over LTE (VoLTE) and other services based on 3GPP's IP multimedia subsystem (IMS) (Ericsson, 2018) (Køien).

In the case of machine-type communications/Internet of Things, the focus was on securing connectivity among IoT devices rather than on hardening access control and data protection. The paradox is that the key vulnerabilities in 5G – from a technical point of view – arise precisely from its outstanding advantages: high throughput, low latency, ability to support massive deployments of IoT devices, virtualization and flexible design based on novel networking concepts such as software-defined networking (SDN), network function virtualization (NFV), multi-access edge computing (MEC), network slicing (NS), and cloud computing (Khan, Kumar, Jayakody, Dushantha, & Liyanage, 2019).

In 2016, the concept of 5G "Vertical Industries" was envisioned by 5G-PPP as the main objective to optimally adapt the operation of the network to the specific needs of a given sector of human activity. The VTF (Vertical Engagement Task Force) has been established under the umbrella of 5G-PPP to coordinate activities related to working with vertical sectors. A *Vertical* is a market comprising a group of companies and customers that are all interconnected around a specific niche. MEC (multi-access edge computing) is a mainstay of such architecture, suitable for URLLC (ultra-reliable and low-latency communication), which supports services requiring short response times, and mMTC (massive machine type communication) for machine-to-machine communications (M2M) intended to hold many devices in a base station. It is based on the split of the Cloud Layer and the introduction of Edge Layer services, where most intelligent functions previously in the former shift to the latter as a way to enable the operation of many IoT devices and to reduce latency (Nowak et al., 2020).

Concerning the 5G standardization, at least four pending matters arise: (a) risks originated in vendor-mandatory/MNO non-mandatory protection schemes in 3GPP TS 33.501 e.g., non-access stratum (NAS) signaling and user plane confidentiality, (b) user plane integrity protection e.g., mandatory UE support of integrity protection and replay protection of user data between the UE and the gNB, but the data rates at which it is supported is different between Release 15 and 16, and it is optional for service providers to use, (c) privacy exceptions for SUPI/IMSI (Subscription Permanent Identifier/ International Mobile Subscriber Identity) where the SUCI (Subscription Concealed Identifier) may use null scheme, thus identity is not protected, and (d) Network and Transport Layer Security schemes (IPsec, TLS) (CSRIC 7, 2020). Verticals such as agribusiness and rural healthcare will need as well a long-term, upgrading security standardization schedule (IMDRF, 2020; MDCG, 2019; Roddy et al., 2019; Kristen, Kloibhofer, Díaz, & Castillejo, 2021).

1.5 Agribusiness and Smart Farming

When currently referring to smart farming, there is an implicit association with a sort of technological revolution comprising scientific advances in a broad variety of disciplines, from genetics to climatology, irrigation, machinery, food and textile supply chains, and manufacturing logistics. However, technological transformations involving the farm are rather a continuous process started long ago, likely in the 19th century with the advent of basic mechanical processes of harvesting and textile production along with steam engines in thresher/harvester machinery. In reality, such continuous tech transformation encompasses operations in and off the farm, because the farm is nowadays a two-way player, the cornerstone of a sophisticated supply chain of food and fiber products and services requiring efficient and secure – inside and outside – processes.

The word agribusiness, first introduced by John Davis and Ray Goldberg in 1955 was defined as follows in their famous 1957 book: "Agribusiness means the sum total of all operations involved in the manufacture and distribution of farm supplies; production operations on the farm; and the storage, processing, and distribution of farm commodities and items made from them. This agribusiness essentially encompasses today the functions which the term agriculture denoted 150 years ago" (Davies & Goldberg, 1957).

The Food and Agriculture Organization (FAO) claims that farmers need to produce 70% more food in the next 30 years to feed the world's growing population. The fast-growing needs of food require new technologies in agriculture to produce more crops in less space and with less human intervention. The increasing use of digital technologies in agriculture and their applications in smart farming is often assumed as a positive driver for environment sustainability, clean production, and reduction of greenhouse gases emissions. However, most developing countries with intensive use of technology in agriculture have experienced as well as a variety of drawbacks and environmental costs, including the loss of biodiversity and forest areas (Winkler, Fuchs, Rounsevell, & Herold, 2021).

In almost all Latin American countries, ICT in agribusiness and particularly, in 5G precision agriculture and livestock traceability is seen as a key factor for quick increases in productivity and exports. Argentina, Brazil, and other large area countries have a long history in digital technologies applied to the agribusiness ecosystem. Precision agriculture in the region started in the mid-1990s due to incoming capital streams, good prices for corn and wheat, and the expansion of sown areas. Despite varying crop prices, expensive fuel and climate disasters with floods, wildfires, and droughts, agriculture machinery manufacturers in South America have been incorporating precision farming tools to existing and new products. For example, in 2019 in Argentina, 90% of planting equipment sold included planter monitors, and 25% also bore variable-rate seeding control. In a recent survey about the adoption of smart agriculture applications in Argentina, 78.2 % out of 1044 respondents said they were already using them, while 66.8% indicated gains in the profitability of the farm as a result of such applications. Satisfied users with 1500 hectares or more accounted for 44.8% of the sample, but there were also 26.4% with less than 300 hectares (Scaramuzza, 2021).

The ultimate barrier for 5G smart farming is (and will be for a while) the geographical coverage of 5G infrastructures, mainly in developing countries. However, according to a recent report, if connectivity is implemented successfully in agriculture, the industry could tack on USD 500 billion in additional value to the global gross domestic product by 2030. Agriculture is one of the seven sectors that, fueled by advanced connectivity, will contribute USD 2–3 trillion in additional value to global GDP over the next decade. By 2030, advanced connectivity infrastructure of some type is expected to cover roughly 80% of the world's rural areas, except Africa, where only a quarter of its area will be covered (Goedde & Revellat, 2020).

Current SF systems are anyway already running in a sort of half-power IoT mode, collecting on-site data with a grid of in-field, low-cost sensor devices deployed throughout large farm lots. In the most basic way, these IoT sensors are scanned from mobile handsets or tablets aboard the tractor/planter/tiller machinery via Bluetooth, although comporting long delays in data processing and high fuel plus manpower costs. An upper arrangement is that of WSN (wireless sensor networks), where each lot in the farm has one or more in-field multi-sensor devices, each consisting of an ultra-low-power microcomputer harnessed to a UHF transceiver, which is connected to the nearby lot sensor(s) and ultimately, to the gateway node in a meshed or mixed mesh/bus architecture (Figure 1.4).

Another scenario is the smart dairy farm (or any livestock farm) where IoT is used for cattle real-time identification, tracking, and individual care by means of RFID tags. This entails a different application because animals move around the farm area, thereby requiring nodes with trilateration or received signal strength (RSS) capabilities. Animal tracking in extensive livestock farms with large areas requires either long-range telemetry transmission or RF proximity systems with a high number of interconnected nodes. Although cattle tracking applications historically were performed on wide areas by using GNSS telemetry (global navigation satellite systems), battery life became a drawback so regular husbandry procedures shifted to RFID localization because barcode ear tag identification of animals is already mandatory in many countries' legislation (Cosby, Manning, Fogarty, & Wilson, 2021).

Again here, the use of low-power wideband area network (LP-WAN) in the 915 MHz band is better in terms of power consumption and distance

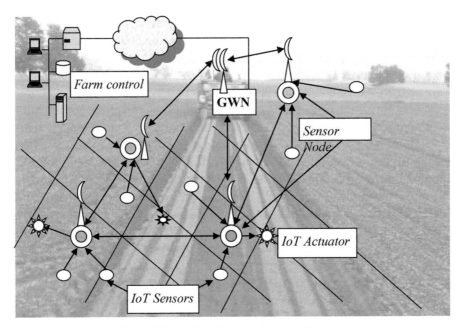

Figure 1.4   IoT architecture for smart farming.

ranges – up to 15 km with a single node –, with data rates up to 200 Kbps. To a large extent, these SF deployments are intended for small-size data applications (position, temperature, soil moisture, pest control, etc.) and consequently, they can run even on older 2G or 3G environments and/or LPWAN such as 3GPP-standardized LTE-M and NB-IOT, or proprietary platforms like LoRA or SigFox. Narrow bandwidth and few latency requirements let them cost-attractive when sensor data is not retrieved too frequently – battery lifespan is up to 15 years in NB-IoT (Akinsolu, Sangodoyin, & Adeyemi, 2021).

A second stage in smart farming was the advent of medium bandwidth, noncritical video or audio sensors, such as photo cameras or slow-motion video monitoring. These systems require more battery power but usually can run over Wi-Fi (2.4 GHz) or UWB (3.1–10.6 GHz) access networks, provided that temporary link outages or interferences do not let the whole network off. A systematic review of scientific papers on IoT in agriculture until 2019 found that WSN (29%) and Wi-Fi (15%) were the most widely used connectivity platforms, followed by Zigbee (10%), RFID (8%), and LoRA (6%). In order to optimize agribusiness for specific applications, it was proposed as one of the verticals, on account of the need for new services with enhanced capabilities and real-time operation in the smart farming niche (Farooq, Riaz, Abid, Umer, & Zikria, 2020; Nowak et al., 2020).

For example, real-time cameras, video streaming cattle recognition, and location tracking or services alike need low latency and high bandwidth, although not still in the mission-critical reliability range. The operation of smart tractors, tillers or planters, surveillance drones and artificial intelligence agribots, among other advanced applications, not only require higher latency and speed parameters but also stringent reliability and hardened security conditions. MEC-enabled functions include fast processing of e.g., video sensor images at the edge and immediate direct feedback to appropriately drive the steering wheel actuator of a smart vehicle. Each sensor within an edge node is connected through a gateway that links edge data with a network, so the sensing grid acts as the front-end edge of the IoT system. The grid gathers measurements of soil and crop moisture, temperature, etc., and sends them to the edge cloud through a gateway node. Data may be classified and filtered before uploading, so as to process at the cloud just the relevant measurements. 5G/URLLC is vital to keep low latency in critical feedback decisions as soon as the processed relevant data is available (Nowak et al., 2020; Ali, Pal, Kumari, Karuppiah, & Conti, 2017; Beavers, 2018).

In the literature, there are extensive research and models referred to cybersecurity and privacy in smart farming, some of them by analyzing the prospects for 5G, besides the abundant studies on the same matter for IoT networks. It has been identified a number of cyberattack models and suggested countermeasures, and classified them according to different criteria (Wu, Yang, Lee, Chu, & Kumar, 2021; Kumar & Sharma, 2021; Yazdinejad et al., 2021; Racovita, 2021; Dorairaju, 2021; Gupta, Abdelsalam, Khorsandroo, & Mittal, 2020; Correa Lima, Lira Figueiredo, Barbieri, & Seki, 2020; Brummer, 2021).

Low-cost sensors, RFID, and other physical devices operating in unlicensed bands are easy targets for attackers using RF jamming or side channel approaches. RF jamming is likely the easiest and cheaper hardware attack, widely utilized against vehicle/home burglar alarms: it consists in masking the RF signal from the IoT sensor with a stronger signal from the nearby attacker, which mimics the legitimate (unprotected) sensing modulation. Side-channel attacks occur when any unauthorized parameter of the implementation of the IoT device is monitored and leaked by a "side" attacker e.g., cache use, power, or processing time variations. The probability of occurrence of security violations at the physical layer is high, particularly in large farms with a lot of legacy IoT devices scattered in completely open and unprotected areas. Once at the network layer, most known threats such as denial-of-service (DoS), man-in-the-middle (MITM), sinkhole, and Sybil attacks are dangerous in the MEC architecture because even when having originated at the bridge level, they can propagate deeper into the entire system (Nowak et al., 2020; Fonyi, 2020; Rugeles, Guillen, & Cardoso, 2021).

IoT sensors work as dummy devices in WSN grids, whose open and vulnerable nature will probably last for years due to low costs, unlicensed RF operation, and low power consumption. The cloud firewalls are unable to see the mesh communications among IoT sensors at the lower level and with the gateway nodes (GWNs) (Figure 1.4). In fact, GWN become the Achilles' heel of SF verticals because of two counteracting needs: they have to filter traffic from many IoT sensors and actuators in quite short times to preserve their low latency features, but anyway they must keep high-level security by processing strong, 2-factor or even 3-factor authentication algorithms. Furthermore, to carry the burden of such processing duty, GWNs require extra power consumption, not available anytime and anywhere inside the farm. In fact, a great share of current research on smart farm IoTs is devoted to lightweight IoT authentication mechanisms, symmetric and asymmetric keys, and quantum cryptography (Wu, Yang, Lee, Chu, & Kumar, 2021; Gupta, Abdelsalam, Khorsandroo, & Mittal, 2020; Wazid, Kumar, Shetty, Gope, & Rodrigues, 2021; Ahmed, 2021; Thakor, Razzaque, & Khandaker, 2021). However, as pointed out by Kristen et al (2021), today's cybersecurity standards and regulations are not sufficient for securing the agriculture domain against new and domain-specific cyberattacks (Kristen, Kloibhofer, Díaz, & Castillejo, 2021).

1.6 Rural Healthcare 5G

One of the certainties endorsed by the 2020 pandemic was the need for a Copernican turn in healthcare access and delivery, where 5G is expected to unleash bioconnectivity by decentralizing hospital services and expanding the provision of timely, high-quality medical care not only at home but also in rural areas and "smart ambulances". In just two months – from February to April 2020 –, the global use of telemedicine grew 78 times, although at the end of the year it has stabilized at a 38-fold increase. In the USA, the demand of telehealth services by consumers grew from 11% in 2019 to 46% in 2020, obviously reflecting a large substitution effect to in-person medical visits (Bestsennyy, Gilbert, Harris, & Rost, 2021).

The sharp increase was attributed to greater willingness to use telehealth applications by both patients and providers, besides changes in regulation and reimbursement policies enabling wider access to virtual consultation and remote medical practices. The global digital healthcare market was valued at near USD 96.5 billion in 2020, of which telemedicine represented some USD 45 billion, and it is expected to grow at a 15.1%

CAGR. Various projections suggest the market will exceed USD 500 billion by 2025, with telemedicine reaching at least USD 175 billion in the same period. For the next five years, 5G's contribution to economic growth will be fairly modest, as telecom companies focus on infrastructure construction and rollout. But starting in 2025, the investments will have an increasingly energizing effect on the global economy, as 5G-enabled applications become more widespread (Chow, 2021; Ericsson, 2021; Bestsennyy, Gilbert, Harris, & Rost, 2021).

5G is expected to impact also in emergency medical services (EMS), even in rural areas. In the UK in 2016, approximately 500,000 ambulance hours were lost due to both hospital transfer and post-transfer preparation of the vehicles. In the USA, each year nearly 10 million rural households receive EMS care. Nationwide, there are 23,272 ambulance agencies, 73% of them serving rural areas. However, rural ambulance agencies are now overwhelmed by long distances and challenging terrain that prolong emergency response and transport times, insufficient payment by insurers to cover costs, changing workforce that has historically relied on volunteers but increasingly must include paid personnel, lack of regional plans to coordinate services, and insufficient state and Federal policy coordination across oversight agencies (Zhai, Xu, Chen, Lu, Wang, & Li, 2020; Mueller, Coburn, Knudson, Lundblad, McBride, & MacKinney, 2021). The response time for emergency medical services (EMS), one of the key performance indicators (KPI) for pre-hospital care to increase the chances of survival should be less than 8 minutes according to the WHO. In the United States, the US EMS Act establishes a difference: 95% of emergency requests should be served within 10 minutes in urban areas but within 30 minutes in rural areas (Cabral et al., 2018; Talukder & Hass, 2021). Most ambulances in the United States and other countries are already equipped with cellular in-vehicle networks supporting computer-aided dispatch, mobile data terminals (MDTs), automated external defibrillators (AEDs), live video streaming, and connected medical devices. A comparison analysis among 4G- and 5G-connected ambulances yielded in the first scenario, 5G unloaded uplink rates nine times better than in a 4G network. In the second scenario, the average measured delay of 5G was 12.88 Mbps, while that of 4G was 76.85 Mbps – six times greater (Ericsson, 2021; Zhai, Xu, Chen, Lu, Wang, & Li, 2020).

In Latin America and the developing world, governments are investing and taking active roles in digitizing their health systems and building the necessary infrastructure to support these initiatives. In Argentina, the CyberHealth project plan includes upgrading hospitals to allow for video-conferencing and connecting 325 healthcare institutions across Argentina to facilitate remote consultations and expertise sharing. Likewise, Brazil has implemented intensive care unit programs (ICU) in rural areas to facilitate access to larger hospitals, where physicians collect and interpret vital signs in real-time and use pan-tilt-zoom (PTZ) cameras to inspect patients visually in the remote location. Brazil has two telemedicine programs: TBR (Telessaúde Brasil Redes), under the aegis of the Ministry of Health, and RUTE (Telemedicine University Network) provided by the Ministry of Science, Technology, and Innovation. Both countries have announced projects to develop health databases over the next years to record patient information from healthcare providers across the country, including consultations, hospitalizations, medications, tests, and vaccines (USTDA, 2020).

Of course, international cooperation has also a presence in the region. For example, since 2019 the World Telehealth Initiative (WTI) has had two programs involving remote consultation in pediatrics in Northern Argentina, serving indigenous communities. Volunteer physicians of the Garrahan Pediatrics Hospital in Buenos Aires can beam the patients via telehealth and bring expert advice avoiding long, costly, and risky flight travels (WTI, 2020).

Although small medical IoT devices e.g., blood pressure, glucose, and pulse rate meters use to offer Bluetooth connectivity to smartphones and personal digital assistance (PDAs) and can be rapidly deployed in rural zones, the 5G era has prompted a closer examination of such integration policies, even at the top government and regulatory levels. Due to cybersecurity and privacy issues, health care will probably become the hardest and most cumbersome service ever targeted for 5G applications, with sound ethical and legal concerns and prone to long and hot debates worldwide.

According to 2021 cybersecurity guidelines from the European Union, healthcare IT systems have two specific conditions that make them different from the rest: (1) patient data is permanent, cannot be changed if privacy is broken (as a credit card number can be), and (2) cyberattacks can become physical and cost human lives (Drougkas, Liveri, Zisi, & Kyranoudi, 2020). In developing countries, the vast majority of medical healthcare ICT devices in rural hospitals and small healthcare facilities are old units, running legacy operating systems below minimum protection levels. Some devices, especially those on the cheaper side, cut a lot of corners in their security in order to cut costs. Because of the low-grade security, attackers like to target these items (Unit 42, 2020). A report has found that 83% of medical imaging devices run end-of-life, unsupported OS, and 57% of IoT devices are vulnerable to medium- or high-severity attacks. For example, devices operating outdated OT protocols like DICOM (Digital Imaging and Communications in Medicine) are exposed to integrity attacks: it was found that attackers could change the header in a packet to replace the image captured by the device with an executable file. As the image was saved, the malware persisted on a network drive. If another DICOM device opens the image, the viewer executes the image and runs the malware. Since DICOM images use to store patient information, antivirus tools are not allowed to scan their locations for privacy reasons. In short, those malware pieces were protected by design (Biamis & Curran).

Moreover, the flood of inexpensive IoT devices for homecare and integrated care poses additional concerns: manufacturers are trying to enlarge their offerings by adding digital services, while clinics and healthcare providers strive to retrofit their out-of-date medical inventory into existing IT infrastructure. Both types of devices were never designed to operate in IoT networks, therefore lacking the required security design to operate in this environment. The original manufacturers of some old medical devices either never intended to provide updates, or stopped doing so (Gleißner, Dotzler, Hartig, Aßmuth, Bulitta, & Hamm, 2021).

A global regulatory framework for cybersecurity of medical IoT devices has been promoted by the International Medical Device Regulators Forum (IMDRF), based on four general principles.

1. Global harmonization: Convergence of efforts at a global level is necessary to ensure that patient safety is maintained while encouraging innovation and allowing timely patient access to safe and effective medical devices.

2. Total product Lifecycle: Risks associated with cybersecurity threats and vulnerabilities should be considered throughout all phases in the life of a medical device, from initial conception to end of support (EOS). To effectively manage the dynamic nature of cybersecurity risk, risk management should be applied throughout the total product life cycle (TPLC) where cybersecurity risk is evaluated and mitigated in the various phases of the TPLC including but not limited to design, manufacturing, testing, and post-market monitoring activities.

3. Shared Responsibility: Medical device cybersecurity is a shared responsibility between stakeholders including manufacturers, healthcare providers, users, regulators, and vulnerability finders.

4. Information Sharing: All responsible stakeholders are encouraged to actively participate in information sharing analysis organizations (ISAOs) to foster collaboration and communication of cybersecurity incidents, threats, and vulnerabilities that may affect the safety,

performance, integrity, and security of the medical devices and the connected healthcare infrastructure.

Medical IoT devices targeted for e.g., 5G applications may contain software, firmware, and programmable logic controllers (e.g., pacemakers) or either exist as software only (e.g., Software as a Medical Device (SaMD)). In the design and manufacturing stages, the IMDRF guidelines have proposed a sample of considerations to be borne by medical device manufacturers, related to seven design principles: secure communications, data protection, device integrity, user authentication, software maintenance, physical access, and reliability/availability, as depicted in Table 1.3. (IMDRF, 2020; MDCG, 2019).

Referring to principle 5 (software maintenance), IEC 62304:2006 and AMD1:2015 (6.2.5 – Medical device software – Software life cycle processes), requires manufacturers to inform users and regulators about any problem in released medical software and how to obtain and install changes.

To reduce the risk of patient harm for vulnerabilities, the national regulator may direct the manufacturer to disable specific functionality of the medical device, accessories, or the supporting ecosystem (e.g., software update servers). Recommended patching methods include: (1) remote update, (2) use administered, and (3) service visit. In order to shed light on 5G expansion-security tradeoffs, it is worth examining South Korea's national plan as an example of current and forthcoming debates about smart healthcare. In 2019, when the South Korean National strategy for 5G was unveiled, digital healthcare was one of the five proposed core services (the other being smart factories, immersive contents, smart cities, and autonomous vehicles). For the mobile telecommunications industry, Korea's plan was a primer and an important caveat: the country reached its first million 5G subscribers in just 69 days after its official launch in April 2019 – and accounting near 17 million by mid-2021, 23 % of all mobile subscriptions (Samsung, 2019; Hong, Ryu, & Lee, 2021).

The Ministry of Science and ICT (MSIT) unveiled in 2019 a USD 6 million project to build a 5G-enabled medical system intended to provide advanced emergency response. It includes a software platform for emergency data integration and analysis, accessed in real-time by ambulances on the go to retrieve and convey patient information (voice, video, bio-signals). Although telemedicine in South Korea is banned by law (Medical Services Act) since 2002, teleconsultation is permitted among patients and healthcare providers. Dentists and doctors (including oriental medicine) can use computers or visual communication systems for remote consultation, but anyway they must diagnose and treat patients in person. There were

Table 1.3 IMDRF guidelines.

Design principles	Proposed design considerations
1. Secure communications	How the device would interface with other devices or networks. May include hardwired connections and/or wireless communications. Examples: WiFi, Ethernet, Bluetooth, USB.
	Design features that validate all inputs (not just external) and take into account communication with devices and environments that only support less secure communication e.g., a device connected to home network or a legacy device.
	How data transfer to and from the device is secured to prevent unauthorized access, modification, or replay. Example: determinate how the communications between devices/systems will authenticate each other; if encryption is required; how unauthorized replay of previously transmitted commands/data will be prevented; and if terminating communication sessions after a pre-defined time is appropriate.
2. Data protection	If safety-related data that is stored on or transferred to/ from the device requires some level of protection such as encryption. Example: passwords should be stored as cryptographically secure hashes.
	If confidentiality control measures are required to protect message control/sequencing fields in communication protocols or to prevent the compromise of cryptographic keying materials.
3. Device integrity	Evaluate the system-level architecture to determine if design features are necessary to ensure data nonrepudiation, e.g., supporting audit logging function.
	Risks to the integrity of the device such as unauthorized modifications to the device software.
	Controls such as anti-malware to prevent viruses, spyware, ransomware, and other forms of malicious code of being executed on the device.
4. User authentication	User access controls that validate who can use the device or allows granting of privileges to different user roles or allow users access in an emergency. Additionally, the same credentials should not be shared across devices and customers. Examples of authentication or access authorization include passwords, hardware keys, or biometrics, or a signal of intent that cannot be produced by another device.

Continued

Table 1.3   Continued.

Design principles	Proposed design considerations
5. Software maintenance	Establish and communicate a process for implementation and deployment of regular updates. How operating system software, third-party software, or open source software will be updated or controlled. The manufacturer should plan how to respond to software updates or outdated operating environments outside of their control (e.g., medical device software running on an unsecure operating system version). How the device will be updated to secure it against newly discovered cybersecurity vulnerabilities. Example: whether updates will require user intervention or be initiated by the device, and how the update can be validated to ensure it has no adverse effect on the safety and performance of the device. What connections will be required to conduct updates and the authenticity of the connection/update through the use of code signing or other similar methods.
6. Physical access	Controls to prevent unauthorized access to the device. Example: controls with physical locks or physically restricting access to ports, or not allowing access with a physical cable without requiring authentication.
7. Reliability and availability	Design features that allow the device to detect, resist, respond, and recover from cybersecurity attacks in order to maintain its essential performance.

contingent relaxations during the pandemic, but personal information and privacy are also strictly protected by law (Personal Information Protection Act). Physicians must obtain prior explicit consent from patients to share any personal information of the subject with third parties – for any purpose – since they are considered sensitive data. Furthermore, the Ministry of Public Health has issued guidelines for the de-identification of patients, e.g., in the case of medical research using technologies such as AI and precision medicine platforms, but there are claims about their complicated and uncertain interpretation (Tursunov, Lenox, & Cleave, 2019).

As described, rural healthcare is another 5G niche market with enormous potential demand but at the same time requiring hard investment efforts in connectivity infrastructure, advanced security-hardened medical equipment, and sound privacy and safety regulations. Furthermore, the 5G network is just an intersection point in the landscape of digital health innovations like machine learning, artificial intelligence (AI), augmented reality (AR), virtual reality (VR), big data analytics, and will be unrealized without them. Therefore, a customized introduction strategy for rural 5G health care must take into account which (advanced) services, applications, and equipment are it supposed to connect, when and where will they be available.

1.7 A Vision of Expansion-Security Tradeoffs

5G is just an enabler for technology applications and infrastructure assets whose ubiquitous access and readiness of use must be thoroughly examined before embarking in expansion ventures. In the developing world, it may be the shortcut to long-lasting upgrades in a variety of vertical industries and national welfare goals. Conversely, it will turn to risky, useless, and expensive luxury wherever no concrete conditions for its fruitful and secure application exist, even with new stakeholders and greenfield rollout strategies (NPN and native Stand-Alone networks) (Aranda, Sacoto-Cabrera, Haro-Mendoza, & Astudillo, 2021; Trakadas et al., 2021). The cybersecurity dimensions requiring intervention may arise from a self-assessment through the various international frameworks, as explained in Section 1.3. At a glance, the rationale of 5G adoption in developing countries relates to four questions: (1) Are there pressing needs in the country that can be profitably addressed by 5G? (2) To what extent (and why) is it the best alternative to fulfill them? (3) Are there vertical industries ready to leverage its unique features and to derive concrete benefits from their early adoption? (4) Is there a cybersecurity and privacy policy and development background suitable to manage the additional risks borne by 5G? (Dimitrievski et al., 2021).

Trends in local infrastructure and costs are key indicators because cloud services need increasing storage and processing capacity with stringent cybersecurity standards. In Latin America, the data center market is expected to cross USD 7.8 billion by 2026, growing at a CAGR of 7.6%. In 2020, Brazil contributed with over 40% of total investment, and now has 45 out of 185 tier-III certified data centers in the region, while the revenue for colocation services grew by over 10% in such period (Arizton Research, 2022).

According to our understanding, the road ahead in 5G expansion worldwide will take the form of a two-speed rollout, where MEC verticals such as smart farming and digital healthcare in customized niche markets will overtake national ecosystems and conventional MNO expansion strategies. Accordingly, in spite of urban-centered earlier 5G forecasts, rural areas in those niches will not have to wait that long. It will ultimately depend on the ability of regulators and concerned stakeholders in solving the puzzle

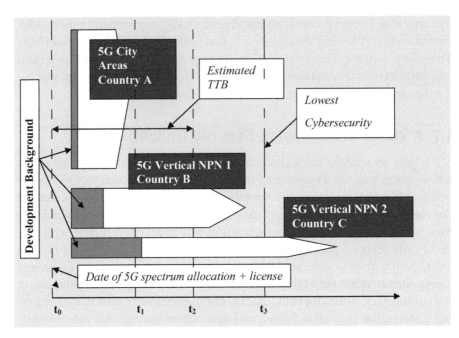

Figure 1.5   Hypothesized scenario-5G expansion-security tradeoffs in developing countries.

of spectrum decisions, new alliances, and suitable applications before the expected rise of cybersecurity issues (Matinmikko-Blue, Yrjölä, Ahokangas, & Hämmäinen, 2021). These mutual agreements will result in local expansion tradeoffs reflecting feasible intersections between investment, scope, and required cybersecurity level in the envisaged 5G project. Spectrum decisions on 5G bands are up to national and regional regulatory agendas, although due to pandemics, in several cases the timelines are delayed. There are countries still in consultation stages, others already have allocated, released, or even granted spectrum chunks through competitive bidding processes (e.g., Chile in early 2021) (Forge & Vu, 2020).

The hypothetic scenario in Figure 1.5 outlines how these deployment tradeoffs may take place in three developing nations (A, B, and C) with a common, spectrum-enabled starting date (t0) and the relative scopes of their proposed use cases stacked in the Y-axis. A sense of the required development effort (CAPEX/OPEX, expertise, management, and support capacity) is represented by the two-sector arrows' internal areas – seemingly equal. Inside each arrow, the development background and the oncoming effort are respectively represented by gray-shaded and white areas. In country A, the MNO has determined that a proposed 5G-NSA urban coverage target with e.g., regular picocell modules (transceiver + RF front-end + panel antennae) will reach neither economic feasibility (estimated time to breakeven, t2) nor the required maturity for 5G cybersecurity, represented by the time to benchmark, t3. However, the upgrade may get feasible if the country enhances its over-all development background –i.e., data centers, infrastructure investment-. Security will remain an issue unless a bold, combined public-private effort achieves significant changes in benchmarking dimensions (Section 1.3). A consortium consisting of agricultural and broadband/MNO cooperatives with agro machinery manufacturers has devised a 5G smart farming project under a vertical non-public network format (NPN 1) in country B. Due to its smaller scope and better development background, this project is feasible whenever the oncoming development effort is sufficient to surpass the TTB (t2).

Nevertheless, the cybersecurity framework self-assessment yields a score value below the acceptable benchmark at (t3), thereby the consortium should improve the cybersecurity background before starting-up, or try another suitable tradeoff between scope, resources, and feasibility.

The third vertical industry case is a 5G-enabled smart healthcare project in Country C, comprising e.g., an EMS corridor to connect smart IoT-ambulances and hospitals along a multi-city route and the rural neighboring zones. A large hospital and two universities have partnered with a new MVNO to undertake the investment on the basis of their background (e-health skilled doctors, digitized patient records, equipped ambulances, etc.). The envisaged development effort focused on a relatively narrow scope is sufficient for rapidly reaching TTB (t2). Country C has enforced a data protection law and basic regulations in digital security, so its international frameworks' scores –albeit not satisfying – have improved in the last years. An internal self-assessment has found cybersecurity weaknesses in the current mobile health service – i.e., unprotected MIoT devices, old OS, e-security training, risk management policies – but there is a suitable plan to appropriately address them within the time span (t3). Note that the planned 5G architecture will comport at least two or three slices to convey services of different priority, bandwidth, and latency: regular high-quality video patient monitoring, vital signs signals, medical records, emergency calls, and ambulance GPS-assisted dispatch operations (Zhai, Xu, Chen, Lu, Wang, & Li, 2020; Mueller, Coburn, Knudson, Lundblad, McBride, & MacKinney, 2021; Roddy et al., 2019).

The hypothesized scenario reflects in part, how decisions may take place in developing countries, on account of the unforeseen 5G rural paradox reported in Section 1.2, with at least three consequences drawn: (a) New stakeholders and alliances can leverage opportunistic features of 5G via NPN or mixed PN/NPN business models, in selected verticals including rural service areas. (b) Time to breakeven is no longer the ultimate constraint in 5G expansion projects; now, cybersecurity maturity is, (c) Tradeoffs among scope, related background, and oncoming development effort – including e-security – will define priority verticals and countries, although cybercrime has no origin or destination boundaries.

1.8 User Privacy and Other Overlooked Issues

Cybersecurity and privacy in rural 5G networks, although not yet a top strain for providers and customers, bear a number of hidden issues and costs that may be as cheap or as expensive as they are assumed or ignored. According to Griffith (2021), the United States has been managing cybersecurity efforts largely "China-centric": warnings and formal restrictions, financial incentives, and discussions of technology standards to maintain Chinese telecommunications vendors out of the country, to the detriment of developing a broader security strategy for 5G networks. The underlying reason is quite simple and its roots in geopolitics: there are no US vendors capable to supply at a global scale, end-to-end integrated 5G networks. Only Huawei, Nokia, Ericsson, and – to some extent, Samsung – have products in the three segments of the 5G architecture (End-user, Core, and RAN), thereby holding a key competitive advantage in every sales contract. Another Chinese company (ZTE) is also striving to enter as a big player, but Huawei largely dominates it with over 30% of total global sales and 600,000 5G base stations provided until 2020. The United States does not have a RAN global scale provider – or a bundle of competitive, smaller suppliers of 5G radio equipment – the most expensive segment with 60–65% of the TCO (Total Cost of Ownership) (Griffith, 2021; Ahmed, 2021).

There are, of course, a number of risk concerns other than untrusted equipment. In - 1.4 we commented on the case of South Korea in current and forthcoming debates on 5G-healthcare privacy and personal data protection. Smart farming has also tough barriers for adoption, arising from technical, socio-cultural, and environmental sides. For example, a study in Ireland has found negative views with respect to the concentration of data expertise, and the further inequity between farmers and agribusiness. Some authors have warned about the "narrow set of values" of designers and engineers referring to the concepts of "good farming" and "good technology" as their data selection choices "privilege large-scale and commodity crop farmers by focusing on agronomic crop data and data mapping unusable to organic growers" (Bronson, 2019). Other researchers argued on the need to incorporate social sustainability into technology trajectories and "a new framework of multi-actor co-innovation" to guide responsible socio-technical transitions (Rose, Wheeler, Winter, Lobley, & Chivers, 2021).

Another similar, social and legal discussion is about the "Right to repair". In the United States, when a farmer buys or leases ICT products, the provider requires him to accept its end-user license agreement (EULA). This license bans the equipment to be fixed by an independent firm or by the owner himself since only the manufacturer can repair it. Moreover, the farmer cannot make any changes or upgrades: the manufacturer has the absolute right to improve it during the whole product lifecycle and to unilaterally establish the date of ending its technical support and maintenance. Then, the underlying business model for most smart farm equipment and software is that of hardware and software as a service (HaaS–SaaS). The Digital Millennium Copyright Act classifies bypassing these protections as a breach of copyright, which may make repairing the technology a criminal offense for all but the manufacturer. Given the distance, downtime, and costs involved in getting e.g., a smart tractor to its nearest provider representative for repairs, "an increasing number of farmers are placing greater value on acquiring older, simpler machines that don't require a computer to fix" (Koebler, 2017; Lumbard, Ahuja, & Snell, 2020).

An overlooked detail about rural 5G is that farm telecommunications comprise not only machinery and devices (IoT) but also people (farmers, rural population, and visitors). A cyberattack intended to steal data on the volume and location of sown lots may open unexpected doors to let the attacker reach sensitive personal information stored in nearby handsets or tablets. Location privacy is at risk to the same extent that some MNOs have disclosed or sold user location information to over-the-top service providers (OTT). An MNO knows in which cell the user is located and can also know the position within the cell, because of the pervasive network data analytics function (NWDAF). By aggregating and AI-processing such location data leakages, eavesdroppers using passive radio receivers or rogue NodeBs (gNBs) may open further vulnerabilities at various layers (Tomasin, Centenaro, Seco-Granados, Roth, & Sezgin, 2021).

Furthermore, such risks may come from a two-way intrusion path. In rural areas of the EU and other industrialized countries, there are socio-cultural barriers precluding the adoption of 5G-based SF and/or the full accomplishment of its benefits due to privacy concerns. Uncertainties regarding data protection and sovereignty became the most frequently cited obstacle to acceptance in the literature (Lieder & Schröter-Schlaack, 2021). A socio-cultural study on the willingness to share farm data over a sample of 880 farmers in Australia revealed the hesitation of producers to let farm-level data generated in their properties to be aggregated and analyzed at scale through SF applications: only 34% respondents perceived themselves as the main beneficiaries of the aggregation on such big data platforms, while 35% and 21% considered respectively, agribusiness and government instead as the main beneficiaries, with the level of willingness to share information depending on who was perceived as benefiting most (Zhang, Heath, McRobert, & Llewellyn, 2021).

There are also "rebound effects", when e.g., the pretended lower consumption of sensor devices does not really drive to real energy savings, but rather it spurs a generalized spreading of devices – in number and/or in density – ultimately leading to more power consumption. Likewise, higher crop yields and/or lower production costs finally translates to lower market prices, thus increasing whole demand for agricultural goods, in particular for renewable resources and bioenergy. Higher demand turns farmers prone to expand sown areas even toward marginal lands where (paradoxically), production is not profitable without SF technologies (Lieder & Schröter-Schlaack, 2021).

Notwithstanding 5G is also a big enabler for green IoT-based agriculture, and this is also an important but overlooked tradeoff since blockchain technology has entered its maturity phase and can be profitably utilized. Organic and DOCG (Designation of Origin Controlled and Guaranteed) products can benefit from guaranteed traceability by authenticating critical parameters in the agrifood chain and in the processing of agricultural products. Its adoption in green agriculture could add significant value, depending on timely and right decisions on hardware, data analytics, maintenance, mobility, infrastructure, data security, and privacy (Kamilaris et al, 2021; Ferrag et al, 2021). In sum, the balance of risks and benefits in smart farming security is a part of a more general tradeoff where climate change and carbon emissions are at odds with changes in land use and the option for capital-intensive, productivity-driven agriculture. Likewise, 5G rural healthcare needs a reconciliation between advantages (technology-driven, time-saving, cost-cutting homecare practices via IoT-diagnostics/monitoring using AI plus big data analyses), and hazards (unknown exposure to cybersecurity or privacy breaches, unauthorized access to patient records and personal data, etc.).

1.9 Conclusion

Along this chapter, the opportunities and risks of a prospective rural 5G expansion have been discussed. Due to historical lacks in capital-intensive network infrastructure, absence of top expertise in cybersecurity, and divergent priorities among MNOs, the pathway toward rural 5G in Latin America and the developing world is likely to last several years. Though, this may be good news since preparedness will get better to cope with the changing nature of cybercrime, and to disentangle the complex systems interdependencies in 5G.

As pointed out by George Sharkov, the new threats become "unknown unknowns"; upgrading cybersecurity requires upgrading of the established since centuries resilience principles of the society to the entirely new maturity level, ("cyber resiliency").

The engineering community has three main challenges with rural 5G security. First, the service-based architecture is a new design where undetected design and implementation flaws /shortcomings may remain unknown for a while. Second, cybersecurity became the new feasibility threshold. However, the suitable tradeoffs between opportunistic but reliable application niches for rural 5G networks are so far, just hypotheses. Third, resilience-by-design will be the new principle for engineering rural 5G networks and for every further research on the matter.

References

Ahmed, A. (2021). Lightweight digital certificate management and efficacious symmetric cryptographic mechanism over industrial Internet of Things. *Sensors*, 2810.

Akinsolu, M., Sangodoyin, A., & Adeyemi, K. (2021). Design considerations and data communication architecture for national animal identification and traceability system in Nigeria. *IST-Africa 2021 Conference Proceedings, IST-Africa Institute/ IIMC.*

Ali, R., Pal, A., Kumari, S., Karuppiah, M., & Conti, M. (2017). A secure user authentication and key-agreement scheme using wireless sensor networks for agriculture monitoring. *Future Generation Computer Systems, 84*, 200 - 2015.

Aranda, J., Sacoto-Cabrera, E., Haro-Mendoza, D., & Astudillo, F. (2021). 5G networks: A review from the perspectives of architecture, business models, cybersecurity, and research developments. *Novasinergia, 4*(1), 6–41.

Arizton Research. (2022). *Latin America Data Center Market Report Scope 2021–2026*. Arizton Research.

Beavers, I. (2018). Intelligence at the edge Part 3: Edge node communication. *Technical Report, 2018. Analog Devices Engineer Zone.*

Bestsennyy, O., Gilbert, G., Harris, A., & Rost, J. (2021). *Telehealth: A quarter-trillion-dollar post-COVID-19 reality?* McKinsey Insights.

Biamis, A., & Curran, K. (n.d.). 5G security and the Internet of Things. In *Security and Organization within IoT and Smart Cities*. CRC Press: USA.

Bronson, K. (2019). Looking through a responsible innovation lens at uneven engagements with digital farming. *NJAS-Wageningen Journal of Life Sciences*, 90 – 91.

Brummer, T. (2021). Cybersecurity in beyond 5G: use cases, current approaches, trends, and challenges. *Communication Systems XIV, Technical Report IFI-2021.02 Chapter 3.*

Cabral, E., Silva Castro, W., Florentino, D., Araujo Vianna, D., daCosta Junior, J., Pires de Souza, R., et al. (2018). Response time in the emergency services. Systematic review. *Acta Cir. Bras., 33*(12), 1110–1121.

Campbell, C. (2019). *What the Chinese Surveillance State Means for the Rest of the World.* Retrieved October 19, 2021, from https://time.com/5735411/china-surveillance-privacy-issues/

Casey, K. (2021). *Edge computing and IoT: How they fit together.* Retrieved October 19, 2021, from https://enterprisersproject.com/article/2021/3/how-edge-computing-and-iot-fit-together

CEA. (2014). Cybersecurity Enhancement Act of 2014 (Enacted January 1, 2021). *Public Law*, 113–274.

Chow, W. (2021). *The global economic impact of 5G" Global Technology, Media and Telecommunications (TMT).* PWC.

Cordeiro, M. (2021, 128). *Brasil ganha primeira antena 5G voltada para o agro no sul do país.* Retrieved from Digital Policy Law: https://digitalpolicylaw.com/sercomtel-realizara-proyecto-piloto-5g-en-areas-rurales-de-brasil/

Correa Lima, G., Lira Figueiredo, F., Barbieri, A., & Seki, J. (2020). Agro 4.0: enabling agriculture digital transformation through IoT. *Revista Ciência Agronômica -.*

Cosby, A., Manning, J., Fogarty, E., & Wilson, C. I. (2021). *Assessing real time tracking technologies to integrate with identification methods and national traceability requirements.* North Sydney: Final Report, Project V.RDA.2005. CQ University Australia. Meat and Livestock Australia Ltd.

Creese, S., Dutton, W., & Esteve-González, P. (2021). The social and cultural shaping of cybersecurity capacity building: a comparative study of nations and regions. *Personal and Ubiquitous Computing, 1–15.*

CSRIC 7. (2020). *Report on recommendations for identifying optional security features that can diminish the effectiveness of 5g security - Communications Security, Reliability and Interoperability.* USA: Working Group 3: Managing Security Risk in Emerging 5G Implementations.

Davies, J., & Goldberg, R. (1957). *A concept of Agribusiness.* Boston USA: Division of Research, Graduate School of Business Administration, Harvard University.

Dimitrievski, A., Filiposka, S., Melero, F. J., Zdravevsvki, E., Lameski, P., Pires, L. M., et al. (2021). Rural healthcare IoT architecture based on low-energy LoRa. *International Journal of Environment Research and Public Health, 18*(2021), 7660.

Dorairaju, G. (2021). *Cyber security in modern agriculture case study: IoT-based insect pest trap system.* MsC. Thesis, Jamk University of Applied Sciences, Finland.

Drougkas, A., Liveri, D., Zisi, A., & Kyranoudi, P. (2020). *Procurement guidelines for cybersecurity in hospitals.* European Union Agency for Cybersecurity (ENISA).

Duffy, C. (2020). *The big differences between 4G and 5G.* Retrieved October 19, 2021, from https://edition.cnn.com/2020/01/17/tech/5g-technical-explainer/index.html

Electoral Commission Ghana. (2021). *2020 Presidential Election Update.* Retrieved October 1, 2021, from https://ec.gov.gh/

Ericsson. (2018). *A guide to 5G network security.* Insight Report 18:000589 Uen Ericsson AB.

Ericsson. (2021). *Edge computing - a must for 5G success.* Retrieved October 1, 2021, from https://www.ericsson.com/en/edge-computing

Ericsson. (2021). *Ericsson Mobility Report - EAB-21:005137 Stockholm, Sweden June 2021.* Ericsson.

Ericsson. (2021). *Mobile Data Traffic Outlook.* Retrieved October 19, 2021, from https://www.ericsson.com/en/reports-and-papers/mobility-report/dataforecasts/mobile-traffic-forecast

ETSI. (2021). *5G security architecture and procedures for 5G System (3GPP TS 33.501 version 16.6.0 Release 16)–Technical Specification TS 133501 V16.6.0 (2021-04).* Retrieved from 5G Americas: https://www.5gamericas.org/wp-content/uploads/2021/01/InDesign-3GPP-Rel-16-17-2021.pdf

Farooq, M., Riaz, S., Abid, A., Umer, T., & Zikria, Y. (2020). Role of IoT technology in agriculture: a systematic literature review. *Electronics, 9*(2), 319.

Fontenla, E. (2016). Cooperativas agropecuarias argentinas: nueva generación de cooperativas. *Serie Documentos No 1; Ediciones CGCyM. Buenos Aires.*

Fonyi, S. (2020). Overview of 5G security and vulnerabilities. *International Conference on Cyber Conflict (CyCon US 2019) Defending Forward (Spring), 1*, pp. 117–134.

Forge, S., & Vu, K. (2020). Forming a 5G strategy for developing countries: a note for policy Makers. *Telecommunication Policy, 44*(7).

Frei, F. X., & Morriss, A. (2020). *Begin With Trust.* Retrieved October 10, 2021, from https://hbr.org/2020/05/begin-with-trust

Gabriel, C., & Kompany, R. (2021). *"Open RAN: ready for prime time? The operators' perspective, Research report, April 2021.* London, UK: Analysys Mason Ltd.

Gamreklidze, E. (2014). Cyber security in developing countries, a digital divide issue. *The Journal of International Communication, 20*(2).

GCSCC. (2021). *Cybersecurity Capacity Maturity Model for Nations (CMM) - 2021 Edition.* Global Cyber Security Capacity Centre Dpt. of Computer Science, University of Oxford, UK.

Gleißner, M., Dotzler, J., Hartig, J., Aßmuth, A., Bulitta, C., & Hamm, S. (2021). IT security of cloud services and IoT devices in healthcare. *Proc. XII International Conference on Cloud Computing, GRIDs, and Virtualization (Cloud Computing 2021).*

Goedde, L., & Revellat, J. (2020). Agriculture's connected future: how technology can yield new growth. *Discussion Paper - McKinsey Global Institute, 2020.*

Goldman, J. G. (2014). *How being watched changes you – without you knowing.* Retrieved October 10, 2021, from https://www.bbc.com/future/article/20140209-being-watched-why-thats-good

Goussal, D. (2017). Rural Broadband in Developing Regions: Alternative Research Agendas for the 5G Era. In K. E. Skouby, I. Williams, & A. Gyamfi, *Handbook on ICT in Developing Countries: 5G Perspective.* Aalborg: River Publishers.

Grand View Research. (2021). *5G Infrastructure Market Size Share & Trends Analysis Report 2021–2028, Report ID: GVR-4-68038-234-1.*

Griffith, M. (2021). Balancing the Promise and the Peril of 5G: The State of Play of the United States. *5G Beyond Borders Workshop, Woodrow Wilson International Center.* USA.

GSMA Intelligence. (2019). *5G In SUb-Saharan Africa: laying the foundations.* Retrieved October 19, 2021, from https://data.gsmaintelligence.com/api-web/v2/research-file-download?id=45121572&file=2796-160719-5G-Africa.pdf

GSMA Intelligence. (2021, January). *The economics of mmWave 5G - GSMA Intelligence.* Retrieved October 12, 2021, from https://data.gsmaintelligence.com/api-web/v2/research-file-download?id=59768858&-file=210121-Economics-of-mmWave.pdf

Gupta, M., Abdelsalam, M., Khorsandroo, S., & Mittal, S. (2020). Security and privacy in smart farming: challenges and opportunities. *IEEE Access, 8* (2020).

HM Government. (2020, August 18). *5G mobile technology: a guide.* Retrieved October 12, 2021, from https://assets.publishing.service. gov.uk/government/uploads/system/uploads/attachment_data/ file/913179/5G_mobile_technology_a_guide.pdf

Hong, E., Ryu, J., & Lee, E. (2021). *Entering the 5g era: lessons from Korea.* Digital Development Global Practice, World Bank.

Horwitz, L. (2021). *How Industrial Edge Fuels Real-Time IoT Processes.* Retrieved October 19, 2021, from https://www.iotworldtoday. com/2021/03/18/how-industrial-edge-fuels-real-time-iot-processes/

Hurel, L. (2021). Cybersecurity in Brazil: an analysis of the national strategy. *Strategic Paper SP-54.*

IDB. (2020). Cybersecurity risks, progress, and the way forward in Latin America and the Caribbean. *Cybersecurity Report.*

IMDRF. (2020). Principles and Practices for Medical Device Cybersecurity. *International Medical Device Regulators Forum, Doc. IMDRF/CYBER WG/N60FINAL:2020.*

Immerman, G. (2021). *The Importance of Edge Computing for the IoT.* Retrieved October 10, 2021, from https://www.machinemetrics.com/ blog/edge-computing-iot

ISO/IEC. (2012). *Information security techniques- Guidelines for cyber- security- ISO/IEC 27032.* Retrieved from International Organization for Standardization (ISO)/International Electrotechnical Commission (IEC): https://www.iso.org/standard/44375.htm

ITU. (2020). *Global Cybersecurity Index 2020 - Measuring commitment to cybersecurity.* Geneve: ITU-D (International Telecommunications Union).

Joy Online. (2021). *Ghana Police Service has 18,000 convicts on database – IGP.* Retrieved October 19, 2021, from https://www.myjoyonline.com/ ghana-police-service-has-18000-convicts-on-database-igp/?param=

Khan, R., Kumar, P., Jayakody, D., Dushantha, N., & Liyanage, M. (2019). A survey on security and privacy of 5G technologies: potential solutions, recent advancements and future directions. *IEEE Communications Surveys & Tutorials, 22*(1), 196–248.

Koebler, J. (2017). *Why American Farmers Are Hacking Their Tractors With Ukrainian Firmware.* Vice Mother board.

Køien, G. M. (n.d.). On threats to the 5G service based architecture. *Wireless Personnal Communications, 119*, 97–116.

Kristen, E., Kloibhofer, R., Díaz, V., & Castillejo, P. (2021). Security Assessment of Agriculture IoT (AIoT) Applications. *Appl. Sci., 11*(13), 5841.

Kumar, P., & Sharma, S. (2021). An empirical evaluation of various digital signature schemes in wireless sensor network. *IETE Technical Review, 2021.*

Lewis, M. (2021). *Telstra pushing 5G into regional and rural Australian communities.* Retrieved from Mobilecorp, Sydney-Australia: https://www.mobilecorp.com.au/blog/telstra-pushing-5g-into-regional-and-rural-australian-communities

Lieder, S., & Schröter-Schlaack, C. (2021). Smart farming technologies in arable farming: Towards a Holistic Assessment of Opportunities and Risks. *Sustainability, 13*(2021), 6783.

Lumbard, K., Ahuja, V., & Snell, M. (2020). Open Agriculture and the Right-to-Repair Community Movement. *MWAIS 2020 Proc.15th Midwest Association for Information Systems Conference, Des Moines, Iowa May 28–29.*

Maquinac. (2021). *La pulverización selectiva se afianza en Argentina.* NVS Desarrollos.

Matinmikko-Blue, M., Yrjölä, S., Ahokangas, P., & Hämmäinen, H. (2021). Analysis of 5G spectrum awarding decisions: How do different countries consider emerging local 5G networks? *23rd Biennial Conference, International Telecommunications Society (ITS) Sweden 21st-23rd June, 2021.* Gothenburg.

MDCG. (2019). Guidance on cybersecurity for medical devices. *Medical Device Coordination Group - Doc. MDCG 2019-16, (EU) 2017/745, 2019.*

Meyerhoff, R. (2019). *Argentina gets smarter about sustainable agriculture.* Forbes.

Monzon, L. (2021). *MTN Announces 5G Roll Out Date in Ghana.* Retrieved October 19, 2021, from https://www.itnewsafrica.com/2021/09/mtn-announces-5g-roll-out-date-in-ghana/

Mueller, K., Coburn, A., Knudson, A., Lundblad, J., McBride, T., & MacKinney, C. (2021). *Characteristics and challenges of rural ambulance agencies–A brief review and policy considerations.* USA: Rural Policy Research Institute (RUPRI).

Nai-Fovino, I., Neisse, R., Hernandez-Ramos, J. L., Polemi, N., PoRuzzante, G., Figwer, M., et al. (2019). A Proposal for a European Cybersecurity Taxonomy. *EUR 29868.*

National Communication Authority. (2021, June 1). *Industry Information - Telecom Subscriptions for March 2021.* Retrieved October 19, 2021, from https://www.nca.org.gh/assets/Industry-Report-March-2021-.pdf

National Communications Authority & Ghana Statistical Service. (2020, March). *Household Survey on ICT in Ghana.* Retrieved October 12, 2021, from https://statsghana.gov.gh/gssmain/fileUpload/pressrelease/Household%20Survey%20on%20ICT%20in%20Ghana%20(Abridged)%20new%20(1).pdf

NIST. (2018). *Framework for improving critical infrastructure cybersecurity, Version 1.1.* USA: National Institute of Standards and Technology.

Nowak, T., Sepczuk, M., Kotulski, Z., Niewolski, W. A., Artych, R., Bocianiak, K., et al. (2020). Verticals in 5G MEC-Use Cases and Security Challenges. *IEEE Access, 9*(2020).

NUKIB. (2018). The Prague proposals-chairman statement on cybersecurity of communication networks in a globally digitalized world. *European Union 5G Security Conference, Prague 3 May 2019- National Cyber and Information Security Agency (NÚKIB).* Czech R.

Racovita, M. (2021). *Industry briefing: cybersecurity for the Internet of Things and Artificial Intelligence in the AgriTech sector.* London, UK: Industry Briefing PETRAS National Centre of Excellence for IoT Systems Cybersecurity.

Roddy, M., Truong, T., Walsh, P., Bado, M., Wu, Y., Healy, M., et al. (2019). 5G Network Slicing for Mission-critical use cases. *2019 IEEE 2nd 5G World Forum (5GWF),* (pp. 409–414).

Rose, D., Wheeler, C., Winter, M., Lobley, M., & Chivers, C. (2021). Agriculture 4.0: Making it work for people, production, and the planet. *Land Use Policy, 100.*

Rugeles, J., Guillen, E., & Cardoso, L. (2021). A technical review of wireless security for the Internet of Things: Software Defined Radio perspective. *Journal of King Saud University - Computer and Information Sciences.*

Russom, P. (2018). *Data Requirements for Machine Learning.* Retrieved October 10, 2021, from https://tdwi.org/articles/2018/09/14/adv-all-data-requirements-for-machine-learning.aspx

Samsung. (2019). 5G Launches in Korea: get a taste of the future. *White Paper.*

Scaramuzza, F. (2021). Adelantos tecnológicos y el futuro inmediato de la agricultura de precisión. *(Conference) INTA (Instituto Nacional de Tecnología Agropecuaria. Manfredi, Córdoba. Biblioteca Fundación CIDETER, Argentina 2021.*

Shimabukuro, I. (2021, 12 5). *Brazilian government inaugurates its first 5G antenna in a rural area of the country.* Retrieved from Olhar Digital: Brazil

Świątkowska, J. (2020). Tackling cybercrime to unleash developing countries' digital potential pathways for prosperity. *Commission Background Paper Series, 33.*

Talukder, A., & Hass, R. (2021). AIoT: AI meets IoT and web in smart health-care. *WebSci'21 Companion, June 21–25, 2021, Virtual Event, UK.*

Thakor, V., Razzaque, M., & Khandaker, M. (2021). Lightweight cryptography algorithms for resource-constrained IoT devices: A review, comparison and research opportunities. *IEEE Access, 9*(2021).

Thales. (2021). *5G technology and networks (speed, use cases, rollout).* Retrieved October 10, 2021, from https://www.thalesgroup.com/en/markets/digital-identity-and-security/mobile/inspired/5G

Tomasin, S., Centenaro, M., Seco-Granados, G., Roth, S., & Sezgin, A. (2021). Location-privacy leakage and integrated solutions for 5G cellular networks and beyond. *Sensors, 21*(2021), 5176.

Trakadas, P., Sarakis, L., Giannopoulos, A., Spantideas, S., Capsalis, N., Gkonis, P., et al. (2021). A cost-efficient 5G non-public network architectural approach: key concepts and enablers, building blocks and potential use cases. *Sensors, 21*(2021), 5578.

Tursunov, R., Lenox, J., & Cleave, J. (2019). *Digital healthcare South Korea. Market intelligence report 2019.* UK: Department for International Trade (DIT) - Intralink Group.

Unit 42. (2020). *2020 Unit 42 IoT Threat Report.* CA USA: Palo Alto Networks, Santa Clara.

USTDA. (2020). *ICT project opportunities in Argentina, Brazil and Paraguay. A resource guide for US industry.* USA: U.S. Trade and Development Agency.

Wazid, M., Kumar, A., Shetty, S., Gope, S., & Rodrigues, J. (2021). Security in 5G-enabled Internet of Things communication: issues, challenges and future research roadmap. *IEEE Access, 9*(2021).

Weissman, C. (1969). Security control in the ADEPT-50 time-sharing system. *Proc. 35th AFIPS Conference*, (pp. 119–133). USA.

Winkler, K., Fuchs, R., Rounsevell, M., & Herold, M. (2021). Global land use changes are four times greater than previously estimated. *Nature Communications, 12*(2501).

WTI. (2020). *Annual Report 2020.* World Tele Health Initiative, Santa Barbara, CA. USA.

Wu, T., Yang, L., Lee, Z., Chu, S., & Kumar, S. (2021). A provably secure three-factor authentication protocol for Wireless Sensor Networks. *Wireless Communications and Mobile Computing.*

Yazdinejad, A., Zolfaghari, B., Azmoodeh, A., Dehghantanha, A., Karimipour, H., Green, A. G., et al. (2021). A review on security of smart farming and precision agriculture: security aspects, attacks, threats and countermeasures. *Applied Sciences, 11*(16), 7518.

Zhai, Y., Xu, X., Chen, B., Lu, H., Wang, Y., & Li, S. S. (2020). 5G Network-enabled smart ambulance: architecture, application, and evaluation. *IEEE Network*.

Zhang, A., Heath, R., McRobert, K., & Llewellyn, R. (2021). Who will benefit from big data? Farmers' perspective on willingness to share farm data. *Journal of Rural Studies*.

2

Cybersecurity Threats to 5G's Critical Infrastructure

Gaurav Meena¹ and Mehul Mahrishi²

¹Central University of Rajasthan, India
²Swami Keshvanand Institute of Technology, India
Email: gaurav.meena@curaj.ac.in

Abstract

The Internet has become an essential element of our everyday lives in the 21st century. These things are as vital to our daily lives as the Internet. We are seeing a rising belief that cyberspace is a battleground for power clashes and possibly even outright war. A country's internal security measures are becoming more dependent on cybersecurity. Concerns about security and a more extensive and complicated attack surface are expected to arise from the greater use of virtualization in 5G networks. When 5G connects billions of interconnected devices, the number of targets and capabilities accessible to spies will grow exponentially. In this study, cybersecurity flaws will be investigated in more detail to provide additional direction and find remedies.

2.1 Introduction

Hackers may target computers, smartphones, and other devices to access private information, change it to hurt others, and more. Threats and attacks are used to describe the potential hazard and risks associated with any of the security infractions listed above (Friedman et al., 2008). It is possible to describe cybersecurity in a variety of ways. According to Cristea (2020), cybersecurity is the ability to defend against malicious assaults on computers, servers, network data, and mobile devices.

Cybersecurity aims to guard against unauthorized access, unauthorized use, and malicious manipulation of sensitive data, including individuals, governments, and businesses. The protection of software, tools, and equipment is only one aspect of cybersecurity; another is the assurance of the privacy and integrity of the data being guarded against various threats and assaults (Fischer, 2005).

Mobile carriers are being forced to undergo a paradigm change in their core technology and radio access networks due to an ever-increasing number of customers and service complexity. As a consequence of this paradigm change, a 5G mobile network has been designed. It is presently being deployed in several industrialized nations such as the United States and the United Kingdom. Least developed countries (LDCs) have recently adopted 4G mobile networks; however, the whole rollout phase has not yet been completed (Rahman et al., 2021).

In information technology, cybersecurity refers to the practice or process of guarding against the loss, alteration, or misuse of data stored on computer systems and networks. There will be more devices linked to the broadband cellular network with fifth generation (5G) technology, allowing for faster downloads and reduced latency, but the flood of new data that these new gadgets will generate must be safeguarded to maintain national security. High capacity, near-zero latency (4–5 millisecond reaction time), high speed up to 10 Gbps, support for a broad variety of applications, 100 times the number of devices that can be accommodated, additional software update possibilities, and pervasive connection are only some of the notable aspects of 5G networks. Such a small number of options come at a cost. The rollout of 5G technology is a massive undertaking, made even more difficult for least developed countries (LDCs). Developing nations having a gross national income per capita of less than $1035 are classified as LDCs by the United Nations (UN). We performed our study based on the UN's list of LDCs (Swarna, 2022).

A new technological generation has been made available to the mobile population in decade-long cycles starting in the '80s. The first-generation mobile network was implemented in the 1980s, the second-generation in the 1990s, and the third- and fourth-generation networks in the 2000s and 2010s. In several wealthy nations, 5G services started rolling out in 2019, and the technology is predicted to be widely available by 2025. However, the difference between 4G and 5G seems to be relatively tiny, especially in LDCs. Even though many wealthy nations began deploying 4G networks as early as 2010, in LDCs, the 4G networks were first established in the decade of late 2020. LDCs' mobile operators will likely bear the brunt of the push to develop 5G networks.

5G, the next generation of wireless mobile technology, provides higher data speeds, lower latency (more excellent responsiveness), and the ability to connect to many devices simultaneously. With these qualities, it is projected

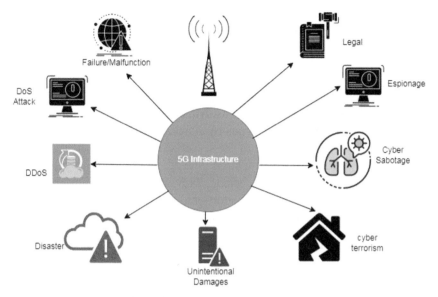

Figure 2.1 5G critical infrastructure.

that robotics, automation, virtual/augmented reality, AI/ML, and machine learning would dramatically influence the world of intelligent devices and applications. 5G will link various parts of society via the network, ranging from key infrastructures (Figure 2.1) such as e-health systems, transportation systems, and electrical grid systems to user environments like smart homes and mobile devices (Agiwal et al.,2016).

Assailants begin by identifying the 5G assets' vulnerabilities and then exploit them by evaluating the attack surfaces. Nefarious acts/abuse, eavesdropping/hijacking/interceptions, intentional and accidental damages, and outages are all examples of cyber–physical security concerns. Cybercrime and physical security risks focus on the first two types of harmful behavior. To steal, change, or destroy the objective; nefarious activity/abuse targets the ICT infrastructures. Eavesdroppers or hijackers can listen, take, or disrupt services through illicit communication channels. People and physical buildings may be harmed or damaged due to intentional or inadvertent damage. Disruptions to service availability and quality fall under the topic of outages. The most significant and severe dangers to 5G are those posed by criminal behavior or misuse (Basnet et al., 2021; Femia et al., 2004; Sauer et al., 2012; Kumar et al., 2019).

This chapter aims to identify probable attack sites and examine security needs and suggestions. For nonspecialists, we want to provide a succinct and thorough instruction.

2.2 Related Studies

2.2.1 Cybersecurity issues in 5G

Denial of service: The primary goal of DoS is to interrupt the availability of 5G and EVSE services. For example, botnets and DDoS may flood network components and base stations with traffic jam, or interfere with the radio frequency, replay, and amplify assaults. Attacks that cause a denial of service are among the most damaging types of attacks. Because of this assault, the legitimate user cannot make use of the server's services. A hacker's ability to connect to the Internet may be impaired if he or she wastes network resources or server resources on the user's behalf. DoS attacks often include a combination of large-scale flooding and additional methods like SYN/UDP/DNS/ICMP and others (Kumawat et al., 2014).

Malicious code: Malicious code injection into the software environment adversely affects system operations, control activities, and operational conditions. A few of the most common types of malwares are viruses, worms, trojans, rogueware, and ransomware.

Exploitation: Hardware and software flaws or vulnerabilities may be found in almost all systems. The attacker may use zero-day exploits, open API, and edge API exploits to access the network and software/hardware.

Abuse: A significant risk of misuse of remote access to the network, authentication/authorization, information leakage, virtualization, and even lawful interception exists because 5G-based EVSE is a highly complex and heterogeneous cyber–physical system with poorly developed administrative coordination and control.

Manipulation: There are several ways an insider or an outsider might get their hands on your network's resources. The possibilities are side-channel attacks, memory scraping, false nodes, rogue MEC gateways, and UICC format exploitation.

The DDoS attack is one of the most effective cyberweapons available. Once in place, it does not wait for a specific server configuration or network situation before attacking or disrupting the target machine's functions. This method does not need a significant expenditure and may result in severe reputational and financial damage to the firm. Ransomware, spyware, adware, worms, Trojans, and botnets are just some of the several sorts of malware that may infect a user's computer, and they all have one thing in common: they are all malicious (turn user PC into a Zombie) (Dahiya et al., 2021).

It is known as "cyber espionage", when an attack on a computer system is carried out to obtain secret or sensitive information or intellectual property to gain an advantage over another firm or government agency. Economic espionage is the purposeful pursuit or acquisition of trade secrets by a foreign power from a domestic enterprise or government body. An industrial spy works for a commercial firm rather than a foreign government to acquire an edge over the competition. Industrial espionage is also referred to as corporate and commercial espionage in certain circles. Intellectual property crime encompasses many actions, including computer espionage and the massive trade in counterfeit goods. The term "industrial espionage" will be used throughout this editorial, but not the phrase "economic espionage". Identifying two competing sets of interests is more significant may be difficult when governmental and business interests overlap (Button, 2020).

In addition, there are always dangers from compromised suppliers, spectrum sensing, data breaches, illegal operations, identity theft/spoofing, signaling storms/frauds, and so on. A few instances of recent cyberattacks occurred in various parts of the world. Government agencies, military and high-tech firms, and economic crimes resulting in losses of more than $1 million are the primary targets of this report.

Apple has been the victim of ransomware attacks. Using a Taiwanese business that manufactures MacBooks and other products for the Cupertino, California-based iPhone manufacturer, Russian hackers launched a $50 million ransomware campaign on the company. The developers of the REvil ransomware have requested a ransom payment from Apple to restore access to the company's critical data. Apple is a well-known victim of the expanding ransomware business, and its data has been encrypted and held hostage. Initially, this included distributing harmful software to computers owned by members of the general population. Any images, documents, or other provided files would be encrypted and rendered unreadable by the program. If the victims pay a ransom, the hackers may agree to provide the decryption key necessary to recover the damaged information.

In December 2015, a hostile cyberattack was responsible for the world's first reported power outage. Hundreds of thousands of people were left without electricity for six hours due to the Black Energy malware, which infected three Ukrainian utility companies. Israel's National Electricity Authority reported that it had been the target of a significant cyberattack on its systems due to the outage that had occurred. As part of an attempt to prevent the spread of a virus, Israel's electricity utility shut down networks.

Insider attacks are a vital cause of concern in today's cyberthreat environment, as 40% of businesses reported as the most damaging. Every organization in 2016 spent $4 million on countering and resolving insider threats. In 2015, 55% of all cyberattacks were the result of insiders (Bradley et al., 2015). However expensive and often occurring, intruder assaults by insiders remain elusive.

According to the alliance, in the first half of 2020, ransomware attacks were responsible for up to 41% of all claims. Palo Alto Networks, a cybersecurity company, expects that in 2020, the average ransom demand would jump from $115,000 to $312,000, on average. Cryptocurrencies are increasingly being used to pay ransoms. For all of us, the $5 billion in 2020 premiums might mean a lesser impact on individual assaults, but the overall effect could be more severe.

2.2.2 Challenges and technical aspects

It is possible to adopt 5G even though it has just a few features and a few potential advantages. The following is a list of some of the most pressing issues.

Aspects of spectrum allocation

It is necessary to have a large quantity of spectrum available in order to achieve faster data speeds and vast network capacity with 5G. For its macro cells, the need calls for a low-frequency band of less than 1 GHz, a mid-frequency band (in the range of 2.3–3.5 GHz) for its micro cells, and a high-frequency band (mmWave in the range of 26–100 GHz) for its small cells.

Costs of spectrum

The rising cost of spectral resources is another issue. More than three times as much as industrialized nations pay for spectrum is shown by recent research to be the case in developing countries. As a result, the cost of spectrum might have a significant impact on the rollout of 5G in different countries (Vaezi et al., 2022).

Inadequacies in infrastructure

One of the most important parts of ensuring 5G coverage and capacity is having access to the necessary infrastructure. Equipment required to support base stations and their interoperability are referred to as "infrastructure". The back-haul element of a 5G network relies heavily on fiber connection to attain huge capacity. In 4G, the majority of base stations do not have fiber connections (Indoria, 2020). Maintaining an optical fiber infrastructure is a costly endeavor that requires a lot of time and effort. As a result, mobile carriers often employ microwave radio spectrum to link back-hauls. It's possible that microwave communication might be hampered by the fact that it needs a direct line-of-sight.

Costs associated with increased network density

Though the network density is poor, even if the spectrum is ample, the feasible network capacity will remain low. The number of base stations spread throughout a region in relation to its population determines the network's density. The larger the density, the more expensive it is. In Latin America, Africa, and India, lower density is a common phenomenon. It is necessary to increase the density of 5G networks to support the cluster of micro cells. This need adds to the difficulty of raising the required funds and constructing sufficient base stations to provide acceptable 5G network coverage (Holma et al., 2020).

Spectrum sharing dynamically

The utilization of cognitive radios is one of the distinguishing characteristics of 5G. Such radios have the ability to discover and utilize available channels in the surrounding area on an opportunistic basis. It is necessary, however, to first develop a strategy for spectrum sharing among mobile carriers; otherwise, prospects that have been identified may not be used because of payment issues. For fear of income losses, regulatory authorities in least developed countries (LDCs) often do not allow spectrum sharing between operators.

2.3 Security of 5G in LDCs

While the 5G network poses a broad variety of security vulnerabilities, least developed countries (LDCs) are woefully underprepared to deal with them. In the past, it has been seen that LDCs were unable to address security breaches in other fields of activity.

The lack of security experts

Because their education systems are not adequately supported to create qualified workers, LDCs are suffering from a significant shortage of competent workers (Granville et al., 2000). Education expenditure accounted for approximately 5.98% of GDP in developed countries in 2012, whereas it accounted for only 3.6% of GDP in least developed countries (Arman et al., 2020). Compared to the previous year, this spending climbed to 7.16% in wealthy nations, while remaining almost unchanged at 3.81% in LDCs (Roser et al., 2016). With such a low level of investment in education, it is impossible to generate effective IT employees as well as security specialists, making the maintenance of emerging technologies like 5G networks even more difficult. Additionally, governments' sub-optimal investment in employee training and poor expenditure in research and development (R&D) are more accountable for this situation. There is no replacement for well-trained security specialists when dealing with the security problems of 5G. Importing professional labor and information technology workers continuously is neither economical nor sustainable.

Infrastructural deficiencies

Hardened infrastructure and proper legal and regulatory framework are necessary to defend the 5G network from security threats. Complex and secure network infrastructures are hindered by substandard hardware and inadequate equipment. LDCs are unable to acquire advanced security technology due to a lack of financial resources.

Inadequate financing for research and development

R&D initiatives for particular industries are absent in least developed countries (LDCs) (Utoikamanu, 2019). The share of government spending on research and development as a percentage of GDP is less than 1% in LDCs. As a result, they lack the necessary expertise and resources to build their own software and hardware, and they must rely on imports from other countries. Because of their lack of area-specific expertise and experience, dealing with the security risks associated with these exogenous technologies is much more difficult for them.

Security education is not being provided

Higher education in industrialized nations includes an extensive spectrum of information technology security education, which is taught at all levels from high school through undergraduate and graduate degrees. For example, the National Initiative for Cybersecurity Education (NICE) was established in the United States in order to improve the country's long-term cybersecurity posture. Several colleges have established cybersecurity programs that are backed up by a highly qualified labor pool of graduates. However, in LDCs, information technology security education and awareness are not part of the academic curriculum. They do not have an information technology and security education program because of a lack of interest, money, and willpower. As reported by (Catota et al., 2019), on average, in Ecuador, which is a moderately developed country, 30% of university academic curriculum offer no course on cybersecurity, 50% offer one course, and only 20% offer two courses on cybersecurity, and students tend to avoid these courses because they are optional in the majority of universities in the country. As a consequence, LDCs are unable to create trained labor to protect a nation's essential infrastructure, such as its 5G network, against cyberattacks.

An unwillingness to adapt to new circumstances

In a study of one of the LDCs, Gambia, Tauray et al. discovered 43 ICT hurdles to development (Touray et al., 2013). People's reluctance to adopt new technologies and resistance to change, say the authors, are major roadblocks to the widespread adoption of any new technology. There are three main obstacles to technology adoption in impoverished nations, according to Mushfoq Mobarak of Yale SOM: information failure, cost, and risk aversion. People are reluctant to adopt new technology because they are unable to see how they might benefit from them, as evidenced by the least developed region of Cambodia. It will not be much different to set up 5G in LDCs, where more than 75% of the population still lives in poverty.

2.4 5G Security Strategies

There should be distinct degrees of concern for 5G assets among the entities mentioned in Section 3 and responsibilities for risk mitigation impacting these assets. Stakeholders must establish methods to reduce their vulnerability to cyber dangers, either separately or together.

- Cybersecurity must be a top priority for suppliers (e.g., respect laws, regulations, standards, certify their products, and ensure quality in their supply chains).

- Operators of telecommunications networks are tasked with identifying and mitigating potential threats and implementing countermeasures to protect their systems and comply with regulatory requirements.

- For service applications to function correctly, both service providers and clients are responsible for implementing, deploying, maintaining, and activating the relevant security features.

- To ensure that operators take necessary precautions to maintain their networks and services' general security and resilience, regulators must ensure that these measures are implemented.

- Accordingly, governments must take the appropriate steps to defend national security interests and enforce conformance programs and independent product testing and evaluations.

- Security assurance specifications and standards are developed and maintained by standardization development organizations (SDOs), who guarantee that best practices and appropriate requirements and measures are in place.

2.5 Conclusion

In conjunction with industry, the government and regulators should develop complete security and resilience criteria for 5G systems and full fiber networks. These standards should be evident, specific, and actionable, allowing the industry to understand exactly what is expected. It is anticipated that implementing the criteria by network operators (and, via them, suppliers) would reduce network security and resilience risks and assure the protection of national security interests. By increasing the bar on security, new security reforms in the telecommunications industry should ensure recognized standards in place and conformance programs to ensure compliance.

References

Agiwal, M., Roy, A., & Saxena, N. (2016). Next-generation 5G wireless networks: A comprehensive survey. *IEEE Communications Surveys & Tutorials*, 18(3), 1617–1655.

Arman, A., Purwandaya, B., & Saefuddin, A. (2020). The Impact of Quality of Education and Higher Education on Economic Growth. *Journal of Economic Education*, 9(1), 64–70.

Basnet, M., & Ali, M. H. (2021). Exploring cybersecurity issues in 5G enabled electric vehicle charging stations with deep learning. arXiv preprint arXiv:2104.08553.

Bradley, N., Alvarez, M., Kuhn, J., & McMillen, D. (2015). *IBM 2015 cyber security intelligence index*. IBM Security.

Button, M. (2020). economic and industrial espionage. *Security Journal*, 33(1), 1–5.

Catota, F. E., Morgan, M. G., & Sicker, D. C. (2019). Cybersecurity education in a developing nation: the Ecuadorian environment. *Journal of Cybersecurity*, 5(1), tyz001.

Cristea, L. M. (2020). Current security threats in the national and international context. *Journal of Accounting and Management Information Systems*, 19(2), 351–378.

Dahiya, A., & Gupta, B. B. (2021). A reputation score policy and Bayesian game theory-based incentivized mechanism for DDoS attacks mitigation and cyber defense. *Future Generation Computer Systems,* 117, 193–204.

Femia, N., Petrone, G., Spagnuolo, G., & Vitelli, M. (2004, June). Optimizing duty-cycle perturbation of P&O MPPT technique. In 2004 *IEEE 35th Annual Power Electronics Specialists Conference (IEEE Cat. No. 04CH37551)* (Vol. 3, pp. 1939–1944). IEEE.

Fischer, E. A. (2005, February). *Creating a national framework for cybersecurity: an analysis of issues and options.* Library Of Congress Washington DC Congressional Research Service.

Friedman, J., & Hoffman, D. V. (2008). Protecting data on mobile devices: A taxonomy of security threats to mobile computing and review of applicable defenses. *Information Knowledge Systems Management,* 7(1–2), 159–180.

Granville, B., Leonard, C., & Manning, J. (2000). *Information technology and developing countries: Potential and obstacles.* In Tokyo Club Meeting, Munich, Germany (pp. 19–20).

Holma, H., Toskala, A., & Nakamura, T. (Eds.). (2020). *5G technology: 3GPP new radio.* John Wiley & Sons.

Indoria, S. (2020). *Deployment of 5G Networks Challenges for Developing Countries.* In ICT Analysis and Applications (pp. 255–262). Springer, Singapore.

Kumar, V., Teja, V. R., Singh, M., & Mishra, S. (2019). PV-based off-grid charging station for electric vehicles. *IFAC-PapersOnLine,* 52(4), 276–281.

Kumawat, H., & Meena, G. (2014, October). Characterization, Detection, and Mitigation of Low-Rate DoS attack. *In Proceedings of the 2014 International Conference on Information and Communication Technology for Competitive Strategies* (pp. 1–5).

Rahman, A., Arabi, S., & Rab, R. (2021). Feasibility and Challenges of 5G Network Deployment in Least Developed Countries (LDC). *Wireless Sensor Network,* 13(1), 1–16.

Roser, M., & Ortiz-Ospina, E. (2016). *Financing education.* Our world in data.

Sauer, K. J., & Roessler, T. (2012, June). Systematic approaches to ensure correct representation of measured multi-irradiance module performance in PV system energy production forecasting software programs. *In 2012 38th IEEE Photovoltaic Specialists Conference (pp. 000703-000709).* IEEE.

Swarna, R. N. (2022). 5G and Next Generation Wireless Network in Bangladesh: Trends, Opportunities, and Challenges. *International Journal of Science and Business,* 9(1), 25–35.

Suomalainen, J., Juhola, A., Shahabuddin, S., Mämmelä, A., & Ahmad, I. (2020). Machine learning threatens 5G security. *IEEE Access*, 8, 190822-190842.

Touray, A., Salminen, A., & Mursu, A. (2013). ICT barriers and critical success factors in developing countries. *The Electronic Journal of Information Systems in Developing Countries*, 56(1), 1–17.

Utoikamanu, F. (2019). Closing the technology gap in least developed countries. *UN Chronicle*, 55(4), 35–38

Vaezi, M., Azari, A., Khosravirad, S. R., Shirvanimoghaddam, M., Azari, M. M., Chasaki, D., & Popovski, P. (2022). Cellular, wide-area, and non-terrestrial IoT: A survey on 5G advances and the road towards 6G. IEEE Communications Surveys & Tutorials.

3

Critical Infrastructure Security: Issues, Challenges and 5G Solutions from an Indian Perspective

Prashant S. Dhotre[1], Shafi Pathan[1] and Nilesh P. Sable[2]

[1]MIT School of Engineering, MIT Arts, Design, and Technology University, Pune, India
[2]BRACT's Vishwakarma Institute of Information Technology, Savitribai Phule Pune University, India
Email: prashant.dhotre@ mituniversity.edu.in;
shafi.pathan@mituniversity.edu.in; drsablenilesh@gmail.com

Abstract

Following deregulation, the private sector has made significant investments in information technology, power, and telecommunications. Many developing nations, including India, are pursuing infrastructure development through private investment. Critical infrastructures (CI) are becoming an essential aspect of the Internet, supporting many of our daily activities i.e., telecommunication services, financial transactions, transportation, water supply, energy usage, and so on. However, the regulatory frameworks' insufficient priority on crisis management and recovery is the reason for worry. This chapter investigates the numerous vulnerabilities and threats that exist in CI today and the issue in CI includes technical, operational, and behavioral view. The challenges and security attacks on CI are presented in this chapter along with the suitable solutions to mitigate the risks. The use of 5G technologies will help in developing strategies and processes to secure CI. The pivotal point is that 5G provides the platform and networks required to execute automation of different sectors. This chapter provides recommendations to secure CI.

3.1 Introduction

Critical infrastructure (CI) refers to assets whose loss would harm a country's security, economy, as well as safety. This infrastructure is a collection of physical structures, facilities, networks, and maybe other assets that offer services that are essential to a community's and society's social and economic functioning (Mohanty et al., 2022). Not only is CI utilized to save lives, but it is also used to reduce physical and commercial losses. Physical, virtual, or networked infrastructure can all be considered CI. Energy, information technology, the Internet, gas, and other instances cause public concern to fit within the above-mentioned areas. The protection of physical and economic development is crucial for developing nations to achieve long-term growth. Considering India is a rapidly urbanizing developing country, maintaining the resilience of CI is essential to the country's long-term progress.

Previous calamities in India have shown a pattern of CI services being affected, resulting in significant economic loss. Extreme weather phenomena such as tropical cyclones, flooding, droughts, and heat waves have seen a major rise in their influence in certain locations. The power sector, according to disaster history, is both the most important for the running of numerous other services (which rely on electricity) and the most usually affected service by natural calamities such as earthquakes, cyclones, and other disasters. As the frequency and severity of natural disasters and cyberattacks grow, the security of such infrastructure must become a top priority in the context of situational management. Disaster control infrastructure to secure CI assets is needed, which is aimed to protect people and assets in hazard-prone areas from the consequences of the hazard. In the event of a disaster, master planning for safeguarding vital infrastructure is required, based on risk assessment, financial statistics, standardization, and quality assurance of equipment or network. The government must identify and plan for natural calamities, cyberattacks, and man-made threats and vulnerabilities that threaten CI.

Given its expanding importance, 5G technology has been designated as CI. 5G will be a super-high-capacity, ultra-high-speed data system with new design standards tailored to energy-elicited systems, lowering operators' operational expenses. Going forward, the importance of 5G and the risk-management measures that the government will need to develop will grow exponentially. In essence, this technology allows complicated networks to communicate with one another adequately. Massive machine-type connectivity, enabled by 5G, can allow climate-smart agriculture, smart cities, and smart homes, as well as augmented and virtual reality capabilities. It uses six ultra-reliable and low-latency connections to power autonomous vehicles, smart grids, remote patient monitoring, telemedicine, and industrial automation, all of which rely on 5G technology's capacity to enable, interconnect, and power the IoT (Atkins & Lawson, 2021). Consider the possible damage that a cybernetwork assault may do. Attacks might cause distorted signals to be transmitted to self-driving vehicles, inaccurate pictures to be delivered to long-distance operations, traffic lights in smart cities to be altered to cause accidents, communications infrastructure for airlines or trains to be disrupted, corporate secrets of competitor enterprises to be filtered out, and false news narratives to be produced. Imagine infecting security infrastructure or financial systems like stock, currency, or commodities markets with malware, all of which would be constituted as acts of war (Foundation, 2022). 5G will play an important role to secure CI in various sectors as mentioned above.

The chapter presents an overview of CI and its classification. Further, this chapter discusses security issues and challenges in CI, the chapter also describes the different attacks and possible solutions. The chapter concludes with recommendations based on individual, process, and technology.

3.2 What Is Critical Infrastructure (CI)?

CI refers to the assets, systems, buildings, networks, and other components of CI that society relies on to maintain national security, economic success, and public health and safety (Güner, 2017). The concept of what defines CI varies slightly per nation. The term "CI" refers to "organizational and physical structures and facilities of such critical importance to a nation's society and economy that their breakdown or deterioration would lead to continuous supply issues, notable instability of public security, or other tremendous implications" (Fraunhofer-Gesellschaft, 2020). It may also be expressed as "whose assets, systems, and networks, whether physical or virtual, are deemed so critical that their incapacitation or destruction would have a crippling effect on security, national economic security, national public health or safety, or any combination thereof" (Alex Morrow, Phil Pitsky, Amit Samani, 2019).

To a considerable extent, CI components are interdependent. Agriculture needs clean water, water purification, and pumps, all of which necessitate power, which may be generated from the water held behind a dam. Interference with transportation networks might prevent critical manufacturing and medical services from receiving supplies. Telecommunications, information technology, and financial, commercial, and government services are all intertwined. Transportation, water, energy, and communications have been designated as lifeline functions, meaning that their reliable operations are so important that their disturbances and perhaps the failure of one of these operations will have a straightforward impact on the security and resilience of CI across different sectors. Because of these links and interdependencies across infrastructure pieces and sectors, the loss of one or more vital function(s) has an immediate impact on the operation or mission in numerous sectors. As a result, every CI often has an emergency response plan in place to deal with any risks to its functioning.

In addition to lifelines, CI provides functionality. Election infrastructure, for example, was designated a subsector of the Government Facilities Sector in 2017 owing to the importance of free and fair democratic elections as a pillar of the American way of life (Security Cybersecurity and Infrastructure, 2020). Working to minimize risk in collaboration with the public and private sector entities in charge of providing this type of important function is a fundamental component of sustaining public trust in the nation's CI. According to various organizations of different nations, classification of CIs is mentioned in Figure 3.1.

- **Communications:** Protection of information systems and networks (for example, the Internet); telecommunications services on a fixed basis; telecommunications services i.e., radio, navigation, satellite, and broadcasting are all services provided by the company.

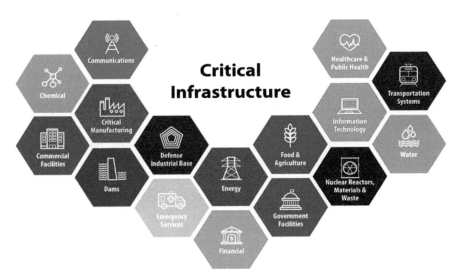

Figure 3.1   Classification of critical infrastructure (Alex Morrow, Phil Pitsky, Amit Samani, 2019).

- **Chemical industry:** Pipelines carrying dangerous goods, as well as the production and storage of hazardous substances.

- **Critical manufacturing:** Various processes involved in the transformation of resources into goods encompass all industrial and transportation operations.

- **Commercial facilities:** Business centers, commercial complexes, sporting venues, and other large-scale meeting places are all examples of large-scale gathering places.

- **Defense industry base:** Facilities to produce military resources (such as weapons, aircraft, and ships), as well as the upkeep of vital national security services (e.g., communications).

- **Dams:** Dams provide people in all parts of the country with a life-sustaining resource. They are a critical component of the nation's infrastructure. Bridges, roads, airports, and other major infrastructure components are all equally important.

- **Energy:** Production, storage, and distribution of energy sources (oil, gas, and electricity) (Alcaraz & Zeadally, 2015).

- **Emergency services:** Emergency services like ambulance, fire brigade, etc. Plays a crucial role in case of emergency.

- **Financial:** Banking, payment services, and government financial assignments are all examples of financial services.

- **Food and agriculture:** Food and agricultural sectors include things like food distribution, safety, and security.

- **Government facilities:** The government provides facilities and functions, as well as armed forces, civil administration services, emergency services, postal and courier services.

- **Nuclear industry:** nuclear materials, production and storage are linked to the nuclear industry.

- **Information technology:** Information technology includes information system and network security, fixed and various telecommunication services.

- **Water:** For achieving water quantity and quality management, the storage and water supply network are critical.

- **Public health and healthcare:** All medical terminologies linked with healthcare and public health include medical and hospital treatment, medications, serums, vaccinations, and pharmaceuticals, as well as bio-laboratories and bioagents.

• **Transportation systems**: Roadways, railways, and air traffic, as well as border surveillance, inland waterway transit, and ocean and short-sea shipping, are all part of the transportation system.

3.3 Security Issues and Challenges in CI

The kind of infrastructure required to provide and maintain regularity in the user's daily life includes high-speed railways, connecting tunnels and bridges, huge highway networks, etc., leads to CI. Subsequently, electricity, water, transportation, etc., depend on the CI (Security Cybersecurity and Infrastructure, 2020). There would be a devastating impact on national public safety, national economy and security or a combination of both if there is the destruction of systems or attack on CI-based systems.

The state and cities depend on the different parameters that form the infrastructure system. The parameter includes critical communications, water supply systems, energy plants, etc. The important concerns or issues around CI include its criticality, time of concern and the attackers, etc. (ANS, 2021). In recent years, attention has been given to CI as it could expose to disturbance by disasters, terrorist attacks, and changes in responsibility and ownership of assets.

The significant issues are espionage, sabotage, and coercion (Mariuz/AAP, 2017). It should be vital to know the kind of threats and their impact on the CI. When it is said to be critical it means not only the national security threats but also the relation between infrastructure and vulnerability, integrity, availability, and the confidentiality of human and environmental systems. The important concern is security, the type of threat, its meaning, and the corresponding loss. As mentioned earlier, the CI include dams (owned by governments), airports (owned by private organizations), irrigation system (owned by the community), or could be electricity provider (owned by the private-public partnership). So, there is a change in the ownership of the infrastructure. Hence, the issue of who is responsible for the change or handling crisis. This could lead to a finger-pointing situation in delivering quality and cost-effective services (Matthew Doran and Alexandra Beech, 2016).

Another issue is the scale of criticality. Looking at the frequent bushfire happening in the Cotter River area could generate contaminated water supply (White et al., 2015) in the eastern part of Australia. It was unprecedented to define the scale of criticality.

The use of the Internet of Things (IoT) is exponential in every sector of CI. The growth of modern techniques and their interconnection with traditional systems is remarkable. The consequences of the use of smart and tiny devices in the CI are leading to more vulnerabilities as compared to isolated systems (Cruz et al., 2016). IoT provides a series of services and offers useful functionalities, but, it is vulnerable and should be managed or separated from the CI to reduce the risk (Zimmerman & Pace, 2018). Hence, the security of critical information is another issue to be addressed.

In India, the use of IoT started a little late as compared to developed countries. However, it has created a canvas to establish and expand device connectivity to provide improved services. The number of devices connected will go from 2.7 billion to 3.2 billion by 2022 (Zimmerman & Pace, 2018).

As the number of smart devices is increasing rapidly, the increase in cyberattack would be seen in several cases observed in recent times like attacks on Cosmos Bank in India (Weinberg, 2021).

3.3.1 Security issues in India

The information flows across the different networks for operational purposes. The kind of services are provided or requested would define the technology and behaviors of the information. The security issues are classified into three main categories technical, operational, and behavioral (Daniel Ani et al., 2016). The classification of the security issues is presented in Figure 3.2.

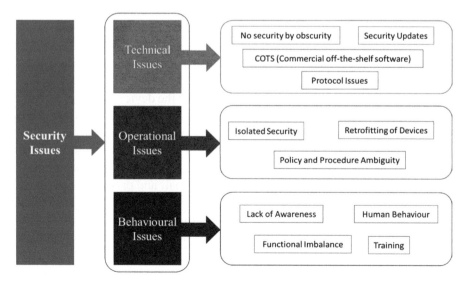

Figure 3.2   Critical infrastructure security issues (Ani et al., 2016).

Technical issues

The security issues in technical aspects are vital in CI management. Security through obscurity is the enforcement of confidentiality and secrecy of the system design architecture. Its objective is to secure a system architecture by intentionally smacking or covering its security flaws. The algorithms are implemented by keeping a password secret. However, it is easily defeated if someone understands and acquires the skills to compromise the system. The issue is how to enhance security by adding obscurity effectively. Another issue is the security updates. It is important to update security patches to ensure updated versions of security mechanisms and services. If the security updates are not installed then there could be potential consequences like the loss of data, damaged software, identity theft, etc. The security updates help in fixing issues related to products, vulnerabilities, etc. The readiness of the system requirements is easily accommodated with the help of COTS. Even though they provide extended solution features, and in a faster way, there are certain issues to be noted that impact the operations of the product or system and the information is gathered and processed. The risk involved could be tampering, poor practices of development, thefts, etc. Based on the type of protocol being used in communication, there are certain issues like protocol attacks.

Operational issues

To provide resilience to the different types of attacks, isolated security is the base of computing systems. The techniques of security isolations depend on the operational approaches and their property analysis. The computing systems work various operations, and the instances are working over the independent machine. However, the issue is to protect those various operations and instances separately. Smart and tiny devices are working together to provide efficient services. Using retrofitting of the devices we could enable existing infrastructure to add on new features or devices. This could be a damage control facility if any vulnerability is observed in the existing infrastructure. On the other side of retrofitting, the issues need to be handled carefully. The issues like assessment and evaluation of the damage, identification of repair techniques, the requirements needed to repair and protect, etc. Another issue in the operation is policy and procedure ambiguity. The security policy allows one to know what to do and what not to do. However, the implementation mechanisms to adhere to the security policy are sometimes ambiguous due to several reasons.

Behavioral issues

For any information system, the important parts are security training and awareness. To improve the process of security compliance and security mechanisms of an organization, security training and awareness play an important role. The behavior of the user depends on the kind of training provided and how much they are aware of. Hence, the main issue is to provide user-centric, product-specific training for developers, operators, etc. Based on the training and the kind of awareness they receive; it is essential to get feedback from the individual. This constructive feedback may help to redefine and provide guidelines to improve security mechanisms training. The issue is how to train and be an aware individual of security in CI.

3.3.2 Security challenges in India

India is the third country in the world where most cyberattacks are observed as per the survey by the National Association of Software and Services Companies (NASSCOM) (Naedele, 2007).

The challenges in providing security to CI include different parameters like technical, operational, financial, etc. In developing countries like India, the important challenge is the lack of sufficient funding for cybersecurity. Considering the available issues of CI and the required security measures, sufficient funding is essentially required. In terms of technical challenges, the importance of computational power is lacking. Also, increased latency is observed during the use of cryptography. The challenge is to improvise the cryptographic techniques to provide confidentiality, availability, and integrity to valuable information in CI. Remote control access is another challenge in CI. The kind of access control mechanism required from a remote location to CI is vital and challenging. At the outset, the challenge from an organizational perspective includes security policymaking and its implementation. The ambiguity in security policies and regulations is another challenge in developing countries like India.

3.4 Analysis/Review of Techniques Used for Security of CI

Due to advances in information technology and the need to improve efficiency, these infrastructures are becoming more automated and connected. This same development has created new risks of mechanical failure, human error, weather, and other environmental causes, as well as physical and cyberattacks. Addressing these weaknesses will require flexible, evolutionary approaches that affect both the private and public sectors and protect domestic as well as international security specific to the infrastructure.

3.4.1 CI security approaches

Ease of collecting large amounts of data related to cybersecurity, information collection on infrastructure about the information technology is affordable as well as manageable. At the same time, it is not technically challenging by a large. Big corporations are aware of cyber threats very much and the first common response of them to the problem is the implementation of infrastructure for data collection. Security measures that can be taken for CI may be listed as

* Identify, analyze, and provide resources for the prevention, detection, disruption, and preparation of threats and risks to CI;

* Prioritize risk reduction efforts, address the physical or operational aspects that make part of the infrastructure vulnerable to exploitation or vulnerability; and

* Minimize the potential effects of events continuously or prepare to reduce them effectively if possible.

3.4.2 Common types of cyberattacks and solutions

With the advances in digital transformation and the complexity of asset networks, it is more important than ever. Key infrastructure providers bring cybersecurity into the 21st century a very important aspect. Successfully tackling this challenge always leads to understanding the risk to CI. Here we are listing the most common types of cyberattacks on CI.

Fraudulent fraud/crime

Identity theft involves an online crime that sends a message, like emails, to the employees of the company which will limit the trusted source. These emails prompt the employee by asking for his or her details and will download malicious software which is hidden as an official file. The receiver assumes the email is from a trusted partner, the employee gives system access. This is harmful to the organization. Stealing sensitive information is a direct form of this practice and is a crime. Cybercriminals invest more time in researching a company and its employees, steal data and use their emails which seem to be more credible.

In the newspaper/channel: In 2015, the hijackers sent a criminal email to steal sensitive information. This information was from the Ukrainian

Parliament. Using a noncomputer program which is attached to the document, remote access was gained to the services of the network – eventually creating a severe power shortage during low temperatures.

Solution:

- Read all emails critically

- Double-check the source of the email before taking action

- Review external links and attachments before clicking

Zero-day attack

This attack occurs when vulnerabilities are detected on a network, software, or computer system and cybercriminals strike before the problem can be resolved. Even if a solution is issued, cybercriminals may take risks. With a zero-day attack, organizations are often unaware of the risk or slowed down in updating or activating their systems.

In the newspaper/channel: The infamous WannaCry ransomware hack has exploited weaknesses in the Microsoft Windows operating system: encryption of user data and ransom required. Computers were at risk when running Windows and had not been updated recently. Twenty-five percent of resource experts were affected by WannaCry or NotPetya attacks (NotPetya also used zero-day vulnerability).

Solution: Avoid zero-day attacks by constantly updating your systems and using episodes.

Brute force attacks/password spraying

Brute force attacks when a criminal enters a large number of simple phrases and common passwords in a single company account. If they are lucky enough to have a not complicated password, they will log in easily. On the other hand, password spraying is when a criminal tries a small number of common passwords in many different accounts – hoping to have at least one company. The account is at risk when passwords are from the pre-selected list.

In the newspaper/channel: The Russian group APT28 (also known as the "Good Bear") used password spraying in targeted attacks in the United States in the past. Although these types of attacks are simple and are usually successful.

Solution: Monitoring this attack is straightforward - create strong passwords for your accounts and enable multi-factor authentication.

Denial-of-Service (DoS) attacks

A denial-of-service attack is on a device or network with traffic. In frustration, the system may crash or – as it handles all criminal applications – is unable to deal with legitimate traffic from employees or customers. DoS attacks cause paralysis, but they are different from other cybercrimes: attackers are not trying to break into the system. Hacktivists and foreign governments may arrange for DoS attacks to cause harm, not to make a profit. The hacktivist opposes the utility business, while the intent of the foreign government controls key U.S. infrastructure. Their motives are not financial.

In the newspaper/channel: Globally, DoS resources attacks have increased sevenfold in the summer of 2020.

Solution: Protecting yourself from DoS attacks requires a strong response system. Keep reading to learn more and more about improving your organization's cybersecurity.

Computer malware

Malicious software, when in your system, may detect connections, steal information, corrupt data, and encrypt files. It comes in many forms and can access your device in many ways. Cybercriminals may still retrieve your files, and your system may be permanently damaged.

In the newspaper/channel: Ransomware was a problem. The BBC reports that approximately 2400 organizations in the United States were targeted by ransomware by 2020, and global costs are estimated at between $42 billion and $170 billion.

Solution: Protecting from malware is not very easy because malware infects your system. A cybercriminal committed to damaging the network, which may even be equipped with resources from outside the government, is extremely dangerous. CI providers must wrap up cybersecurity security in their organization. Continue reading to learn more about preventing cyberattacks on CI.

3.4.3 Recommendations

Generally, the security goals of confidentiality, integrity, and availability (CIA) are applicable to IT systems. Along with the CIA, it is important to consider reliability, performance, risk-free and safety should always be followed. Efficient security management, security risk assessment, and proactive measures are vital, and these must be embedded in the processes, technology and the people should be sensitized. Therefore, this chapter proposes the following recommendations:

- Individual Oriented: Providing the right kind of training and sensitizing individuals about cybersecurity is essential and must be done at regular intervals. Another recommendation includes best practices inculcated about cybersecurity. Security is not a one-time training; upskilling programs are required, and an individual's skills need to be enriched by training and awareness programs.

- Process Oriented: Understanding risk and its assessment, the important recommendation is to define the security process. This could include a strong password policy, threat analysis and associate risk assessment, and defining standard operating procedures. The process should include regular penetration testing of CI and its common parameters. Creating a standard repository of knowledge gathered from different countries about research in CI security.

- Technology Oriented: Once the individual is trained and the standard process is in its place, the next recommendation is the technology and its use. Defining defense strategies is vital and should look into 5G technologies. The technology should focus on active and passive attacks on 5G system components like core network, mobile network, access network, user equipment, and external network. The technology should also look into the coordination of the components with the 4G systems.

3.5 Conclusion

The inevitable influence of CI is observed in different countries including developing countries like India. The cyberspace that connects every aspect of CI is common and vulnerable to various attacks. CI security breaches have different impacts on the national security of India. In this chapter, the CI, its issues and challenges are discussed in detail. Further, this chapter presents common attacks on CI which need to be addressed in the present time. Also, this chapter proposes a solution to the mentioned attacks that include procedures and methods from the perspective of individuals, technology, and process. The use of 5G technologies could be one of the ways to secure CI. The future of CI and its security is important from the view of developing countries. Hence, it is important to understand the importance of CI facilities, associated risk categories, and executing assessment and proactive measures. The privacy and security of CI should include 5G technologies in the countries like India. The integration of CI and cyberspace need to be connected securely and must be coordinated well to avoid serious attacks in future.

References

Alcaraz, C., & Zeadally, S. (2015). Critical infrastructure protection: Requirements and challenges for the 21st century. *International Journal of Critical Infrastructure Protection, 8*, 53–66. https://doi.org/10.1016/j.ijcip.2014.12.002

Alex Morrow, Phil Pitsky, Amit Samani, L. C. A. H. (2019). *Protection of Critical Infrastructure*. Joint Air Power Competence Centre. https://www.japcc.org/c-uas-protection-of-critical-infrastructure

Ani, U. P. D., He, H. (Mary), & Tiwari, A. (2016). Review of cybersecurity issues in industrial critical infrastructure: manufacturing in perspective. *Https://Doi.Org/10.1080/23742917.2016.1252211, 1*(1), 32–74. https://doi.org/10.1080/23742917.2016.1252211

ANS. (2021). *Australian National Security*. ANS Press. https://www.nationalsecurity.gov.au/

Atkins, S., & Lawson, C. (2021). An Improvised Patchwork: Success and Failure in Cybersecurity Policy for Critical Infrastructure. *Public Administration Review, 81*(5), 847–861. https://doi.org/10.1111/puar.13322

Cruz, T., Rosa, L., Proenca, J., Maglaras, L., Aubigny, M., Lev, L., Jiang, J., & Simões, P. (2016). A Cybersecurity Detection Framework for Supervisory Control and Data Acquisition Systems. *IEEE Transactions on Industrial Informatics, 12*(6), 2236–2246. https://doi.org/10.1109/TII.2016.2599841

Daniel Ani, U. P., He, H., & Tiwari, A. (2016). Journal of Cyber Security Technology Review of cybersecurity issues in industrial critical infrastructure: manufacturing in perspective Review of cybersecurity issues in industrial critical infrastructure: manufacturing in perspective. *Journal of Cyber Security Technology, 1*(1), 32–74. https://doi.org/10.1080/23742917.2016.1252211

Foundation, O. R. (2022). *Cybersecurity and Internet Governance*. ORF. https://www.orfonline.org/programme/tech-and-media/cybersecurity-and-internet-governance/

Fraunhofer-Gesellschaft. (2020). *Safeguarding the power supply in the event of a major outage*. Critical Infrastructure Resilience. https://www.fraunhofer.de/en/press/research-news/2020/august/safeguarding-the-power-supply-in-the-event-of-a-major-outage.html

Güner, S. (2017). Operational Efficiency and Service Quality Analysis in Public Transportation Systems. *Journal of Transportation and Logistics*, 2(2). https://doi.org/10.22532/jtl.358727

Mariuz/AAP, D. (2017). *What's critical about critical infrastructure?* The Conversation. https://theconversation.com/whats-critical-about-critical-infrastructure-73849

Matthew Doran and Alexandra Beech. (2016). *SA weather: State-wide power loss raises "serious questions", Josh Frydenberg says By political reporters Matthew Doran and Alexandra Beech*. Abc.Net.Au. https://www.abc.net.au/news/2016-09-28/sa-weather-serious-questions-must-be-answered-frydenberg-says/7886262?nw=0&r=Gallery

Mohanty, S., Dabral, A., Chatterjee, R., & Shaw, R. (2022). Shelter management during pandemics: lessons from cascading risks of cyclones and COVID-19. *International Journal of Disaster Resilience in the Built Environment*, 13(1). https://doi.org/10.1108/IJDRBE-09-2020-0103

Naedele, M. (2007). Addressing IT security for critical control systems. *Proceedings of the Annual Hawaii International Conference on System Sciences, February 2007*. https://doi.org/10.1109/HICSS.2007.48

Security Cybersecurity and Infrastructure, S. (2020). *Critical Infrastructure Sectors*. https://www.cisa.gov/critical-infrastructure-sectors

Weinberg, A. (2021). *Analysis of top 11 cyber attacks on critical infrastructure*. FirstPoint. https://www.firstpoint-mg.com/blog/analysis-of-top-11-cyber-attackson-critical-infrastructure

White, I., Wade, A., Worthy, M., Mueller, N., Daniell, T., & Wasson, R. (2015). The vulnerability of water supply catchments to bushfires: impacts of the January 2003 wildfires on the Australian Capital Territory. *Https://Doi.Org/10.1080/13241583.2006.11465291*, 10(2), 179–194. https://doi.org/10.1080/13241583.2006.11465291

Zimmerman, T., & Pace, B. (2018). *IoT solutions can't be trusted and must be separated from the enterprise network to reduce risk*. https://www.gartner.com/en/documents/3874933

4

OpenRAN and Security in a Developing Country Context

Roslyn Layton, PhD

Communications, Media and Information Technologies, Department of
Electronic Systems, Aalborg University, Denmark
Email: roslyn@layton.dk

Abstract

The marketing and public policy development of open radio access networks
or OpenRAN appears to exceed the academic and empirical investigation of
the technology. The Economic Times of India declares, "Open RAN is the
future of mobile networks. Here's why", (Taneja, 2021) promising to reduce
capital and operational expenditure for radio access networks (RAN) by as
much as half. Despite the brisk promotion in the trade and policy press, there
are but a handful of peer-reviewed articles on OpenRAN.[1] Half a dozen were
written by Ph.D. candidate Leonardo Bonati at Northeastern University.[2]
Earlier described as "software-defined ran architecture via virtualization"
(Yang, Li, Jin, Su, & Zeng, 2013), OpenRAN appears to be a catchall term for
general purpose, vendor neutral hardware; open interfaces; and open-source
software (Telecom Infra Project, 2021). Indeed, telecom analyst Roger Enter
observed, "There are many different flavors of open and even different flavors
of open RAN" (Fletcher, 2021). Moreover, the US-based OpenRAN Policy
Coalition and Deloitte assert the technology's security advantages (Open

---

[1] Aalborg University library search on OpenRAN keywords, December 1, 2021,
https://kbdk-aub.primo.exlibrisgroup.com/discovery/search?query=any,contains,open-
ran&tab=Everything&search_scope=MyInst_and_CI&sortby=date_d&vid=45KBDK_
AUB:AUB&facet=tlevel,include,peer_reviewed&offset=10&came_from=pagination_1_2.

[2] Review of papers by "Leonardo Bonati" in Google Scholar, December 1, 2021, https://
scholar.google.com/citations?user=jZMr9VMAAAAJ&hl=en.

RAN Policy Coalition, 2021) and that these can be leveraged in developing countries with the additional benefit of lower cost (Essing, 2021).

The enthusiasm for nascent technology is interesting and warrants investigation. As such, this article attempts to use a logical, reasoning approach to evaluate OpenRAN claims on security and developing countries and provide questions to help test the proposition of whether and to what degree OpenRAN improves security over regular RAN, and similarly, whether OpenRAN is a compelling proposition for mobile operators in developing countries.

4.1 Introduction

The total global value of the equipment for radio access networks (RAN) is roughly \$35 billion annually (Valuates Reports, 2020). That figure compares to a total of \$170 for capital expenditure for mobile networks globally. In 2019, the world's mobile operators earned just over \$1 trillion and spent roughly \$35 billion on RAN equipment, some 3% of revenue (GSMA, 2021).

The industrial drive for OpenRAN comes in part from mobile operators wanting to reduce cost and improve profitability. OpenRAN can be seen in part as a strategy to commoditize equipment vendors the way that unbundling was proposed as a solution to "open" up telecom networks.

AT&T, China Mobile, Deutsche Telekom, NTT DoCoMo, and Orange formed the O-RAN Alliance in Germany (O-RAN Alliance, 2021). Its purpose is to define the specifications for the technologies which build on top of 5G technical standards from the 3rd Generation Partnership Project (3GPP) which develops the technical specifications for 5G networks, including the suite of security technology specifications (3GPP, 2021). Importantly OpenRAN is not a standard but a specification, a set of requirements that could be fulfilled by various activities and products. A standard, on the other hand, is a formal document that establishes uniform criteria, techniques, and practices.

OpenRAN has been promoted by industry, the trade press, and policy advocates as well as high-ranking government officials. Promoting OpenRAN around the world is the official policy of the US Presidential Administration including the Departments of State and Commerce (FCC, 2021).

This is bolstered by proposals for significant subsidies to OpenRAN firms in the millions, if not billions, of dollars (Morris, 2020). Support is further advanced by a \$500 million grant for OpenRAN development appropriated under the 2021 National Defense Authorization Act (NDAA) and managed by the National Telecommunications Information Agency (NTIA). However, as evidenced by a proposed amendment to the 2022 NDAA, the security assessments of the technology by US authorities are limited but potentially forthcoming). As the amendment necessitates, the assessment of "national security implications" of OpenRAN would come after subsidies are distributed, a provision which would task the Departments of State and Commerce to prepare a report (Spanberger, 2021). Brazil's Ministry of Science, Technology and Innovation pledged 30 million reals (US $5.79 million) for OpenRAN among efforts in other countries (BNamericas, 2021).

Jean-Christophe Plantin of the London School of Economics describes OpenRAN as a geopolitical strategy for governments and industrial actors to leverage the ambiguous definition of openness in an effort to expand control over mobile wireless networks (Plantin, 2021). This could explain the global efforts of US policymakers and US companies at meetings with foreign governments to discuss OpenRAN, for example, at a closed door security policy event in Prague in November 2021[3] and meetings in the United States with international participants (FCC, 2020).

In any event, some valuable applications of OpenRAN have been described in the trade press. These include small cells in certain urban areas depending on the spectrum and transmission rules; "neutral host" networks for multiple operators to use when serving an enterprise customer; and RAN Intelligent Controller (RIC) software that serves as a platform for network applications adapting to performance needs in real time (Morris, 2021). These appear to be a series of one-off applications; whether they can constitute a full end-to-end network seems a stretch and may be indicative of marketing spin. As such, the marketing communications on OpenRAN may represent at least in part the gap between hype and reality. Like making a Tesla from items picked up at Walmart, OpenRAN marketing heralds building a 5G network with an inexpensive, disruptive mix-and-match components overlain with intelligent software.

OpenRAN probably has value in specific situations and application for certain market segments, once the network has been built with regular RAN equipment from a variety of vendors. This means that OpenRAN is not a substitute for the basic building blocks of equipment from regular RAN providers. Operators must still purchase antennas, basebands, remote radios, small cells, macro cells, phase shifters, and so on. The OpenRAN add-ons would follow for certain situations and applications.

---

[3] "Prague 5G Security Conference," 5G Prague Security Conference (blog), November 30, 2021, https://www.prague5gsecurityconference.com/services/.

4.2 OpenRAN and Security

The literature on the security OpenRAN is limited, but three studies merit attention. One is published by the European Commission;" another is a white paper by the OpenRAN Policy Coalition; the other, a report commissioned by Germany's Federal Office for Information Security.

The report "Cybersecurity of Open Radio Access Networks" prepared by EU Member States with the support of the European Commission and ENISA, the EU Agency for Cybersecurity, explains that threat surfaces and vulnerabilities expand in Open RAN functions and interfaces because of an increased number of suppliers, components, and some data processing (e.g. real-time location data of users connected to the network). For example mix-and-match fronthaul interfaces could be exploited to carry out denial-of-service attacks, interception or tampering. This can compromise network availability, confidentiality and/or integrity.

The security risks of OpenRAN include (1) misconfiguration of networks by integrating components from different suppliers; (2) lack of maturity of OpenRAN leading to dissimilar and poorer network design; (3) low product quality from increased number of components from different suppliers which may prioritize security differently; (4) reduced control for MNOs as they must provide access to more suppliers; (5) state interference through the 5G supply chain; (6) increased opportunity for organised crime groups (OCGs) to enter and disrupt networks with more interfaces and suppliers; (7) lack of clarity on whether OpenRAN provides adequate power supply (including backup); (8) vulnerability of virtualised Multi-Access Edge Computing (MEC) applications to attacks from connected terminals, such as IoT devices; (9) expanded threat surface and vulnerabilities in Open RAN functions and interfaces with increased suppliers, components and interfaces; (10) network fault management complexity; (11) new risks in resource sharing as different network functions are virtualised and running on the same hardware.

The report warns of new or increased dependency on cloud service/ infrastructure providers. For example, as virtualisation and cloud services grow, there is a risk of mobile network operators becoming dependent on a small number of cloud service/infrastructure providers, which could lead to supplier lock-in. In addition, networks relying on the same cloud provider could exacerbate vulnerabilities.

The report highlights that Open RAN presents new opportunities, but this also entails new risks. For some opportunities to be realized, the Open RAN interfaces need to be mature, robust, and standardized. Their specifications must be developed with standards organizations requirements like accessibility to information and transparency. Moreover, standards must be accepted by the industrial and regulatory ecosystems. Deploying Open RAN presupposes that network operators and regulators will audit these networks and systems to ensure compliance, which is not certain. The potential use of open source in Open RAN may also require visibility to software sourcing and codes.

OpenRAN Policy Coalition published a four-page paper "Open RAN Security in 5G", which outlines "why Open RAN is not less secure than traditional RAN and can in fact bring security advantages" (Open RAN Policy Coalition, 2021). The OpenRAN Policy Coalition is an advocacy organization of 60 tech companies led by a former NTIA executive and a part-time attorney on loan from Venable, a leading law firm retained by US tech companies.

The first three pages of the white paper focus on the security benefits of 5G overall which represents an improvement of the previous generations of mobile wireless standards. These include: (1) distributed denial of service (DDoS) detection and mitigation, (2) stronger encryption, (3) improved security protocols for roaming, (4) "zero trust" enhancements for core network architectures and application programming interfaces (APIs) which require verification from everything to which they connect, (5) cloud security with assurance and compliance requirements, software signing, signature verification, and vulnerability management processes, and (6) network slicing, the segmentation of the network to allow grouping and separation for sensitive functions.

The white paper asserts that security can be a challenge in a disaggregated or "open" environment, noting "generic security risks such as malware, botnets, and other forms of attacks that are potential risks regardless of the underlying architecture". The paper asserts that OpenRAN will complement and build upon the major security enhancements of 5G. This expressed value proposition appears to be *distributed security analytics* which is described as "direct access to more data about network performance because the components are disaggregated and connected through open interfaces, allowing visibility to potential security problems earlier". It suggests that these analytics could provide more granular data to functions which were previously hidden by vendors. "Since security platform vendors typically provide native support for standard protocols and interfaces, the operator can integrate new security platforms without implementing custom adaptors for vendor-proprietary protocols and interfaces".

It appears that the information is in the system, but it is not presented in the preferred piecemeal format. Essentially OpenRAN's security value proposition is an improved formatting and detailing of information, not necessarily a technical or engineering enhancement which would constitute a new standard. The white paper asserts that OpenRAN allows analytics to be obtained presumably from each component or interface in the system whereas such information is ostensibly obscured or aggregated in regular RAN systems. It is not clear whether mobile operators have demanded improved security analytics, nor is it evident that distributed security analytics is something which established vendors could not supply. However, if the need is to have individual component level information and this is the value of OpenRAN, then it is odd why OpenRAN vendors have not developed this as a technical standard. However novel such an enhancement would be, it does not seem that this capability is as important as the security standards of 5G itself—the incremental innovations of technology which enable better threat detection, encryption, zero trust, cloud security, and network slicing.

A study of the security of O-RAN Alliance specifications was commissioned by Federal Office for Information Security (BSI) in Germany and performed by the Barkhausen Institute (Köpsell, Ruzhanskiy, Hecker, & Stachorra, 2022). The study performs a risk analysis of different actors (the attacker, a 5G network end user, the 5G network operator, and the nation-state), offering best and worst-case scenarios. One scenario is that the attacker uses the open interface, now vulnerable by a jamming attack of spoofed "intelligent" radio signals, to gain control over many IoT devices in the 5G network, and then uses them to conduct a distributed attack. It concludes that OpenRAN specifications are not defined sufficiently to "security by design" principles.

Some related assertions about OpenRAN deserve mention.

4.2.1 OpenRAN and reduced reliance on a single vendor

OpenRAN proponents suggest that OpenRAN improves security because of an imagined future of multiple OpenRAN providers with interoperable components. This is suggested as a superior choice than the status quo of purported availability of mere 1–2 network providers.

However, the market for cellular network infrastructure equipment consists of more than Huawei, Ericsson, Nokia, and ZTE. While there has been significant consolidation of the infrastructure equipment industry in the last two decades, many new firms have entered the industry, notably Samsung.

The 5G network equipment vendor market already has many vendors and segments. Omdia details more than a dozen full-service providers with additional providers in segments for antennas, basebands, remote radios, small cells, macro cells, phase shifters and so on (Omdia, 2020).

This idea that there are not many vendors for 5G equipment is likely a talking point created by Huawei to deter restrictions on its equipment (Owen, 2021).

If anything, it appears that the restrictions on Huawei have opened the door to new equipment vendors which could not compete because of Huawei's purported predatory pricing and anti-competitive tactics. For example, Samsung has quickly gained market share in 5G network equipment and now supplies the United States, Australia, and other countries (Seung-yoon, Sung-yong, & Hyo-jin, 2020).

A test of whether the market is competitive is whether and to what degree a firm can price above its marginal cost. In general, the profit margin in the infrastructure equipment market is thin (this is another reason for consolidation). With Moore's Law, the equipment today is more efficient than in the past. A mobile operator may pay the same or less for a given base station which can do more. In general, vendors have been improving the capacity of their products, even though the a la carte price may be the same. This evolution reflects global competition and the standards development process. The extent to which the vendor can earn more may reflect superior marketing, bundled packaging, cross-sells, service agreements, and so on.

Technological change drives competition in the infrastructure market, not the number of players (Aghion, Bloom, Blundell, Griffith, & Howitt, 2005). Just as the mobile market has evolved from 2G in the 1990s to 5G today, the type and number of equipment providers has changed with technology. Many equipment providers did not last the 2G and 3G generations. If an operator does not upgrade its network to the latest technology, it will lose market position. Similarly, if an infrastructure provider does not innovate its equipment, it will lose customers.

If the assertion that the number of firms mattered, then we would expect equipment prices to increase when Huawei and ZTE were restricted in the United States. This did not happen. Instead, prices held constant (Dano, 2021).

Alternatively, OpenRAN advocates have proffered that vendor-proprietary environments are an "attack surface" whereas OpenRAN would allow mix and match for the best in breed security solutions (Wenger, 2020). However operators have been mixing and matching for years. Indeed, vendors have learned from their customers and incorporated the preferred features into their products.

If anything, OpenRAN itself may be an attack surface as the US Government Accountability Office (GAO) notes in its report "5G Wireless Capabilities and Challenges for an Evolving Network".

The developing Open Radio Access Network Alliance may be the source of another potential vulnerability in 5G architecture. This alliance promotes competition and specialization among a variety of 5G component and software providers, similar to how modern computers use specialized vendors for memory, processing, and software. While this initiative strives to improve performance and reduce costs, the attack surface of the network expands considerably (Gao@100, 2020).

There is no empirical proof that OpenRAN improves security over products from regulator RAN products. Whether OpenRAN is an improvement, will also depend on the costs for the mobile operator to vet not one, but dozens, if not potentially hundreds, of OpenRAN vendors. The time and cost to perform this review would seem to be multiplied by the number of vendors.

Neil McRae, Managing Director and Chief Architect at British Telecom (BT) describes his view (TelecomTV, 2021). McRae explains that when he took his job, he inherited a network portfolio with 50 vendors. He subsequently reduced it to four vendors and saved £1 billion in 3 years. He observed that too many vendors not only increased cost, it increased complexity. He is wary of notions of "open architectures" which require managing portfolios of 5–50 vendors. He noted that vendor reduction increased shareholder value and that he would pursue the same strategy again.

Multi-vendor network environments are not inherently or empirically more secure than networks with one or a handful of regular RAN vendors.

4.2.2 OpenRAN and Huawei

In the United States, the push for OpenRAN has been proffered as a solution to reduce reliance on companies like Huawei and ZTE, companies which are alleged to be vulnerable to intrusion by the People's Republic of China (PRC), whether through technical means like control channels, backdoors, or kill switches, or through government practices of surveillance, espionage, and sabotage (Fletcher, 2021).

As noted earlier, OpenRAN is coordinated globally by the O-RAN Alliance. A founding member is state-owned and license-denied in the USA (Federal Communications Commission, 2019) China Mobile. The alliance also includes 44 PRC companies, which comprise the second largest group of member companies after the United States (O-RAN Alliance, 2021). At least seven of these PRC actors face US Entity List restriction or FCC operating license denial (Federal Communications Commission, 2021) because of the alleged risks they present to national security. US authorities have restricted O-RAN Alliance Members China Mobile, China Telecom, China Unicom,

ZTE, Inspur, Phytium, and Kindroid. Nokia paused its participation in the O-RAN Alliance, noting that US members could violate US law by cooperating with the restricted PRC firms (Cerulus, 2021).

Theoretically, if Huawei and ZTE are not involved in OpenRAN, then the purchase of OpenRAN could reduce reliance on them. However the tradeoff appears to be the integration with new PRC firms in OpenRAN, many of which present security risks of their own. Given the PRC's stated civil-military fusion strategy, the threat of intrusion is likely omnipresent with all PRC technology (Stone & Wood, 2020).

4.3 OpenRAN and Open Source

Open-source software has certain benefits and applications; these include lower software costs, scaling, and other benefits for certain applications (enterprisersproject.com, 2015). Open-source proponents have suggested that open source software is more secure because the code is open for all to see (dwheeler.com, 2021). While some see open source as secure because the code is accessible by all, making vulnerabilities easy to identify, the same benefit makes the vulnerability exploitable and enables reverse engineering for additional compromise (dwheeler.com, 2021).

However with OpenRAN, it appears that Huawei cannot be avoided because it is the lead contributor to Kubernetes, the most-used OpenRAN software platform. A memorandum of understanding between Deutsche Telekom, Telefonica, TIM, Vodafone, and Orange notes that OpenRAN should be built on top of Kubernetes (O-RAN Alliance, 2021). Though Kubernetes was started in 2014 as a Google project, today it is housed in the Cloud Native Computing Foundation, an offshoot of the Linux Foundation, perhaps the world's largest open-source organization. By late 2017, Huawei had gained a seat on the Kubernetes Steering Committee and claims to be the fifth-biggest contributor of lines of software code proposed and accepted to Kubernetes (Huawei, 2017).

Moreover, many developers in the open-source community have a reputation for de-prioritizing security. According to the "Report on the 2020 FOSS Contributor Survey" from The Linux Foundation & The Laboratory for Innovation Science at Harvard, the open-source community spends very little time responding to security issues (an average of 2.27% of their total contribution time) and reports that it does not desire to increase this investment significantly (Linux Foundation, 2020).

Open-source software, whether in OpenRAN or classic RAN, does not necessarily make a network inherently more secure.

4.4 OpenRAN and Developing Countries

OpenRAN has been purported to be a "lifeline" for developing countries because it can significantly reduce capital and operating expenditures (Clark, 2021). For example, the notion is proffered that operators could buy packages of virtualized software instead of hardware. However, the network data still needs a minimum set of hardware on which to run and whose cost is not insignificant.

However, OpenRAN is only available on 4G and 5G, and even that product development is unlikely to be ready until 2025 (Kapko, 2021).

Many parts of the world still use 2G or 3G networks. Millions of subscribers in the Middle East, Africa, Latin America, India, and other part on Asia still use phones that only support 2G and 3G. The most widely used network technologies in Africa today are 2G and 3G. Many developing countries are behind on the rollout 4G. Some parts of Africa and Latin America have acceptable 4G coverage on sites that support 2G/3G and 4G in one base station, but the number of countries in developing markets that have rolled out 5G is still quite limited. In South Africa alone, Vodafone internal data estimates that 6 million 2G devices are sold every year, and about 14 million customers are still using 2G phones (Gilbert, 2021).

GSMA predicts that in 2025, there will still be a substantial 2G and 3G penetration in several regions, particularly the Middle East and Africa with 46% still at 2G and 3G in 2025 (GSMA, 2021). See the following table.

To enable OpenRAN today in an developing country would likely require the mobile operator to have two set of parallel base stations, for example, one set running 2G/3G and another 4G/5G. Having two networks increases

Table 4.1   Penetration of 2G and 3G networks in developing countries (GSMA, 2021).

2025	2G	3G	2G and 3G
Asia Pacific	7%	14%	21%
Armenia, Azerbaijan, Belarus, Georgia, Kazakhstan, Kyrgyzstan, Moldova, Russia, Tajikistan, Turkmenistan, Ukraine, and Uzbekistan	7%	17%	24%
Europe	1%	7%	8%
Greater China	0%	0%	0%
Latin America	5%	21%	26%
Middle East and North Africa (Algeria, Bahrain, Egypt, Iran, Iraq, Israel, Jordan, Kuwait, Lebanon, Libya, Morocco, Oman, Qatar, Saudi Arabia, Syria, Tunisia, United Arab Emirates and Yemen)	10%	36%	46%
North America	1%	6%	7%
sub-Saharan Africa	12%	58%	70%

not only equipment and software costs, but also cost for site rental, energy, spectrum, and sales & marketing for the new network, etc.

Thus the business case for OpenRAN in developing countries appears less favorable than in developed countries.

Similarly, some have promoted OpenRAN as economic development and industrial policy, suggesting it as a way for developing countries to enter the global IT industry (ETTelecom.com, 2021) (Carreño, 2021). Curiously, many of these calls are coupled with operator strategies to keep Huawei equipment in place because OpenRAN will not be ready until 2025 or later (Davies, 2020). Policymakers have also pursued subsidies and other financial incentives to support local OpenRAN startups which may design the equipment in their respective country but manufacture in China. Unfortunately, production in China and with Chinese partners could compromise security, as the Supermicro case demonstrates (Bloomberg.com, 2021).

4.4.1 What is driving OpenRAN hype?

Leading telecom analyst Iain Morris of Light Reading suggests that OpenRAN, which may have legitimate applications, has become victim to religiosity when people question the degree to which it is financially or practically feasible.

"Open RAN has become so riven by feuding geopolitical interests that assessing any claim about the technology on its own merits is virtually impossible. The only sensible question to ask is whether it is or could be the best way to construct a mobile network. Does it make build and operation easier or cheaper? Will it spur innovation? Unfortunately, fanaticism and hostility are crowding out objective analysis.... before Open RAN was a rational debate about improving interoperability between vendors, introducing virtualization into the RAN and rearchitecting networks. Almost overnight, it became the answer to the Huawei conundrum, America's best opportunity to produce rivals to Ericsson and Nokia, a crutch for protectionists, an elixir for operators wounded by their clash with Big Tech" (Morris, 2021).

Essentially Morris suggests that OpenRAN is proffered to solve a political problem that the United States lacks homegrown 5G infrastructure providers. The implication is that the United States, building on its software and IT prowess, peddles an opportunistic narrative to "innovate out" of the Huawei problem with OpenRAN. However, the premise of this problem is not necessarily true; Nokia and Ericsson have the following assets in the United States: manufacturing, tens of thousands of employees, their largest customers, and major shareholders. They may be more American than they are given credit. The United States already has incumbent 4G/5G leadership because its platform companies Google, Apple, Facebook, Amazon, Microsoft, Netflix, etc., take the lion's share of world's market for Internet content, applications, cloud, and advertising. Indeed, these companies are already a leg up in 5G as American consumers "cut the cord" and switch to cellular wireless broadband.

4.5 Conclusion

The chapter analyzes claims about OpenRAN's superiority to regular RAN for security and its suitability for developing countries. It uncovers that the security value proposition for OpenRAN is distributed security analytics. However, it is not evident that this feature is unique to OpenRAN, thus, there appears to be little to no proof that OpenRAN is more secure than regular RAN. The new German study raises many security concerns about OpenRAN. In any event, there are few peer-reviewed studies, so the assertion that OpenRAN is as good or better for security is unfounded.

The preponderance of 2G and 3G networks in developing countries would seem to be the stumbling block for OpenRAN which is not backward compatible to 2G and 3G. Deploying OpenRAN technology in developing countries requires mobile operators to run parallel simultaneous networks; one for existing customers on 2G/3G and another for 4G/5G network which is compatible with OpenRAN.

The article described different geopolitical contours of OpenRAN, for example, the outsized role of the American and Chinese governments and the interplay of other nations. These relations are complex and worthy of their own study. In any event, geopolitics also plays a role in the OpenRAN discussion and may have something to do with the enthusiastic marketing pronouncement of greater security and lower cost for developing countries.

References

O-RAN Alliance. (2021, December 1). *List of Members*. Retrieved from O-RAN Alliance: https://www.o-ran.org/membership.

Spanberger. (2021, September 12). *"Amendment to Rules Committee Print 117-13 Offered by Ms. Spanberger of Virgina." US House of Representatives*. Retrieved from house.gov: https://amendments-rules.house.gov/amendments/SPANVA_062_ORAN%20NDAA%20Amendment_xml210913212731618.pdf.

3GPP. (2021, November 26). *3GPP Specification Series: 33series*. Retrieved from 3GPP.org: https://www.3gpp.org/DynaReport/33-series.htm

Aghion, P., Bloom, N., Blundell, R., Griffith, R., & Howitt, P. (2005). Competition and Innovation: An Inverted-U Relationship. *The Quarterly Journal of Economics, 120*(2), 701–728.

Bloomberg.Com. (2021, November 26). *The Long Hack: How China Exploited a U.S. Tech Supplier*. Retrieved from Bloomberg.Com: https://www.bloomberg.com/features/2021-supermicro/

BNamericas. (2021, August 28). *Brazil to Finance Open RAN Projects*. Retrieved from BNamericas.com: https://www.bnamericas.com/en/news/brazil-to-finance-open-ran-projects

Carreño, I. (2021, November 26). *Operadores brasileños exploran OpenRAN para desarrollar 5G y mantener equipos de Huawei | DPLNews*. Retrieved from digitalpolicylaw.com: https://digitalpolicylaw.com/operadores-brasilenos-exploran-openran-para-desarrollar-5g-y-mantener-equipos-de-huawei/.

Cerulus, L. (2021, August 27). *Nokia Pauses 5G Project Due to Fear of US Penalties*. Retrieved from POLITICO: https://www.politico.eu/article/nokia-pauses-5g-open-ran-work-due-to-fear-of-us-penalties

Clark, R. (2021, June 18). *Open RAN a Lifeline for Emerging Market Telcos*. Retrieved from Light Reading: https://www.lightreading.com/asia/open-ran-lifeline-for-emerging-market-telcos/d/d-id/770317.

Dano, M. (2021, April 28). *Here's how much a 5G wireless network really costs*. Retrieved from Lightreading.com: https://www.lightreading.com/open-ran/heres-how-much-5g-wireless-network-really-costs/d/d-id/769114

Davies, J. (2020, August 6). *Vodafone Switches on UK's First OpenRAN Site*. Retrieved from telecoms.com: https://telecoms.com/505928/vodafone-switches-on-uks-first-openran-site

dwheeler.com. (2021). *Is Open Source Good for Security?* Retrieved from dwheeler.com: https://dwheeler.com/secure-programs/Secure-Programs-HOWTO/open-source-security.html

Enterprisersproject.com. (2015). *8 Advantages of Using Open Source in the Enterprise*. Retrieved from enterprisersproject.com: https://enterprisersproject.com/article/2015/1/top-advantages-open-source-offers-over-proprietary-solutions.

Essing, H. M. (2021, November 26). *Mobile Networks Are Going Virtual—and Open*. Retrieved from Deloitte Insights: https://www2.deloitte.com/us/en/insights/industry/technology/open-ran-deployment-rural.html

ETTelecom.com. (2021, November 26). *Airtel's Dramatic Strategy Shift: Developing Local 5G Gear Ecosystem via Own R&D and US, Japanese Partners - ET Telecom*. Retrieved from ETTelecom.com: https://telecom.economictimes.indiatimes.com/news/airtels-dramatic-strategy-shift-developing-local-5g-gear-local-ecosystem-via-own-rd-and-us-japanese-partners/78774004.

European Commission. (2022, May 11) "Cybersecurity of Open Radio Access Networks." Retrieved from https://digital-strategy.ec.europa.eu/en/library/cybersecurity-open-radio-access-networks.

FCC. (2020, February 20). *Forum on 5G Open Radio Access Networks*. Retrieved from FCC: https://www.fcc.gov/news-events/events/forum-5g-virtual-radio-access-networks

FCC. (2021, July 16). *NTIA*. Retrieved from Promoting the Deployment of 5G Open Radio Access Networks: https://www.ntia.doc.gov/files/ntia/publications/ntia_comments_-_open_ran_noi_gn_21-63_7.16.21.pdf

Federal Communications Commission. (2019, May 9). *FCC Denies China Mobile Telecom Services Application*. Retrieved from FCC: https://www.fcc.gov/document/fcc-denies-china-mobile-telecom-services-application

Federal Communications Commission. (2021, November 2). *China Telecom Americas Order on Revocation and Termination*. Retrieved from FCC.gov: https://www.fcc.gov/document/china-telecom-americas-order-revocation-and-termination

Fletcher, B. (2021, June 9). *Dish, Open RAN Coalition Cheer Senate Tech Bill Passage*. Retrieved from Fierce Wireless: https://www.fiercewire-less.com/tech/dish-open-ran-coalition-cheer-senate-tech-bill-passage

Fletcher, B. (2021, August 20). *Mavenir Swaps out Triangle's Huawei Gear for Open RAN | Fierce Wireless*. Retrieved from Fierce Wireless: https://www.fiercewireless.com/tech/mavenir-swaps-out-triangle-s-huawei-gear-for-open-ran

Gao@100. (2020, November 24). *5G Wireless: Capabilities and Challenges for an Evolving Network*. Retrieved from GAO.gov: https://www.gao.gov/products/gao-21-26sp.

Gilbert, P. (2021, November 26). *Expensive 4G Devices Are Keeping 2G Going*. Retrieved from Connecting Africa: http://www.connectingafrica.com/author.asp?section_id=761&doc_id=772591

GSMA. (2021). *The Mobile Economy," GSMA, 2021*. Retrieved from GSMA: https://www.gsma.com/mobileeconomy/

Huawei. (2017, November 3). *"Huawei Acquires a Seat in the Kubernetes Steering Committee*. Retrieved from Huawei.

Kapko, M. (2021, November). *Is Open RAN Too Little, Too Late for 5G?* Retrieved from SDxCentral: https://www.sdxcentral.com/articles/news/is-open-ran-too-little-too-late-for-5g/2021/11/

Köpsell, S., Ruzhanskiy, A., Hecker, A., & Stachorra, D. (2022, February 21). *Open-RAN Risikoanalyse 5GRANR.* Retrieved from Bundesmt Fur Sicherheit in der Informationstechnik: https://www.bsi.bund.de/SharedDocs/Downloads/DE/BSI/Publikationen/Studien/5G/5GRAN-Risikoanalyse.pdf;jsessionid=5BF4A2F6CB78F0172E-A934B3D089FB94.internet081?__blob=publicationFile&v=9

Linux Foundation. (2020). *2020 FOSS Contributor Survey.* Retrieved from Linux Foundation: https://www.linuxfoundation.org/tools/foss-contributor-survey-2020/

Morris, I. (2020, 15 1). *https://www.lightreading.com/5g/us-senators-propose-more-than-$1b-for-open-ran-to-fight-huawei/d/d-id/756844.* Retrieved from Lightreading.com: https://www.lightreading.com/5g/us-senators-propose-more-than-$1b-for-open-ran-to-fight-huawei/d/d-id/756844.

Morris, I. (2021, December 11). *BT Takes Aim at Open RAN Myths.* Retrieved from Light Reading: https://www.lightreading.com/open-ran/bt-takes-aim-at-open-ran-myths/d/d-id/773471

Morris, I. (2021, November 24). *Nokia Caught in Telco Schism over Open RAN.* Retrieved from Light Reading: https://www.lightreading.com/open-ran/nokia-caught-in-telco-schism-over-open-ran/a/d-id/773721.

Omdia. (2020). *5G: Managing Component-Level Risks for Commercial Success.* Retrieved from electronica.de: ttps://exhibitors.electronica.de/download/1121_11_5_2789_5_1_803/electronica-virtual_5g-white-paperpdf.pdf.

Open RAN Policy Coalition. (2021, September 28). *How Open RAN Can Bring Security Advantages.* Retrieved from https://www.openranpolicy.org/: https://www.openranpolicy.org/how-open-ran-can-bring-security-advantages/

Open RAN Policy Coalition. (2021, September 28). *Open RAN Policy Coalition.* Retrieved from How Open RAN Can Bring Security Advantages: https://www.openranpolicy.org/how-open-ran-can-bring-security-advantages/

O-RAN Alliance. (2021, December 1). *About O-RAN Alliance.* Retrieved from O-RAN Alliance: https://www.o-ran.org/about

O-RAN Alliance. (2021, June). *O-RAN Ecosystem.* Retrieved from O-RAN Alliance: https://www.o-ran.org/ecosystem.

Owen, G. (2021). *The Race to Open RAN Is a Marathon, Not a Sprint.* Retrieved from Huawei: https://www.huawei.com/en/technology-insights/publications/huawei-tech/89/open-ran-is-a-marathon-not-a-sprint.

Plantin, J. C. (2021). The Geopolitical Hijacking of Open Networking: The Case of Open RAN. *European Journal of Communication, 36*(4), 404–417.

Seung-yoon, L., Sung-yong, H., & Hyo-jin, K. (2020, September 8). *Samsung Elec's Network Equipment Market Share to Jump Thanks to Verizon Deal.* Retrieved from Pulse News: https://pulsenews.co.kr/view.php?year=2020&no=926638

Stone, A., & Wood, P. (2020, June 12). *China's Military-Civil Fusion Strategy.* Retrieved from China Aerospace Studies Institute: https://www.airuniversity.af.edu/Portals/10/CASI/documents/Research/Other-Topics/CASI_China_Military_Civil_Fusion_Strategy.pdf.

Taneja, M. (2021, August 7). *Open RAN Is the Future of Mobile Networks. Here's Why.* Retrieved from The Economic Times: https://economic-times.indiatimes.com/tech/tech-bytes/open-ran-is-the-future-of-mobile-networks-heres-why/articleshow/85136051.cms?from=mdr.

Telecom Infra Project. (2021, November 26). *OpenRAN.* Retrieved from Telecom Infra Project: https://telecominfraproject.com/openran/

TelecomTV. (2021). TelecomTV, Open Telco The After Show Live Day 1, 2021. Retrieved from https://www.youtube.com/watch?v=YZ3PHHW_tis

Valuates Reports. (2020, December 04). *Radio Access Network Market Size USD 44.78 Billion by 2026 at CAGR 11.3% | Valuates Reports.* Retrieved from Prnewswire.com: https://www.prnewswire.com/in/news-releases/radio-access-network-market-size-usd-44-78-billion-by-2026-at-cagr-11-3-valuates-reports-850863705.html

Wenger, E. (2020, September 22). *Security in Open RAN Networks.* Retrieved from CISCO: https://blogs.cisco.com/gov/security-in-open-ran-networks

Yang, M., Li, Y., Jin, D., Su, L., & Zeng, L. (2013). OpenRAN: A Software-Defined RAN Architecture via Virtualization. *Proceedings of the ACM SIGCOMM 2013 conference on SIGCOMM, Volume: 43.*

5

Deployment of 5G in Emerging Economies: Cybersecurity Challenges and Potentials for Ghana

Henoch Kwabena-Adade[1] and Ezer Osei Yeboah-Boateng[2*]

[1]Accra Institute of Technology (AIT),
[2]National Communications Authority (NCA), Ghana,

* Corresponding Author
Grace Dzifa Kwabena-Adade
Graduate School, Ghana Communication Technology University,
Email: henochx@gmail.com; ezer.yeboah-boateng@nca.org.gh;
gdkadade@gmail.com

Abstract

Mobile communications networks facilitate the connection of mobile portable devices to operators' networks. The nature of transmission uses unguided electromagnetic media or radio signals, which provides the needed flexibility of mobility for the devices to roam across the coverage areas or cell sites. Based on propagation and transmission technologies, there have been a couple of generations. The latest fifth generation (5G) comes as an upgrade of previous generations, with faster data transmission rates, more features and functionalities. 5G technologies plan to utilize three (3) electromagnetic spectrums; with each spectrum providing unique opportunities for mobile network operators and consumers. For example, it anticipates enabling IoT services, easy propagation through walls, high-speed data rates, with wider coverage areas, etc. Indeed, 5G promises to break the limitations of current mobile communications networks, in respect of bandwidth, latency, and speed. The above notwithstanding, 5G is also anticipated to be accompanied with some concerns, such as possibilities of applications, cybersecurity vulnerabilities emanating due to extensive usage of machine-to-machine communications, and high dependencies of organizations and productivity on the 5G technology. Some pitfalls will surface as usage of 5G grows. There could be challenges dealing with patching unknown vulnerabilities. Edge computing, empowered by 5G, may introduce data leakages as data processing is carried at the data sources. Lack of regulation, standards and compliances in emerging economies, may facilitate cybercriminals tampering with sensors to steal data, as well as the possibility of compromising IoT cameras for homes and associated privacy concerns.

5.1 Introduction

Cellular network is a wireless mobile network. A mobile network is a communication network with wireless last mile, that is devices connect to mobile network via unguided media such as electromagnetic wave transmission, but the backend of the network is usually fiber optic which is a form of guided electromagnetic signals transmission. The unguided nature of the transmission media provides the flexibility of mobility where the connected devices can roam or move freely within the coverage area also known as cells. Because cellular networks do not require the use of cables to connect device to the network, it is relatively easy to enroll large number of devices and it is flexible to upgrade when capacity is exceeded. Cellular network is therefore all about mobile devices which are primarily handheld, or pocket sized or small electronic devices though fixed and larger devices could also be configured to connect to a mobile network.

There have been different generations of cellular networks; from first generation analog networks to the new 5th Generation digital network also

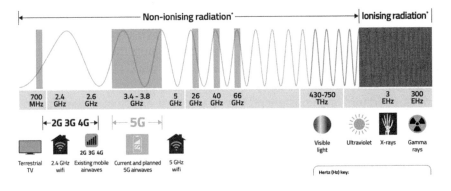

Figure 5.1   Electromagnetic spectrum (HM Government, 2020).

known as 5G. Newer generations are upgrades of the previous generations with faster data transmission rates, more features, and functionalities. All the generations operate on the basic fundamentals of wireless communication.

There is nothing fundamentally different about the physical character-istics of the radio signals that will be produced by 5G compared to previous technologies like 3G and 4G thus, they all use the radio spectrum of the elec-tromagnetic spectrum (HM Government, 2020).

5.2 5G and How It Works

According to the US Congressional research service report with the title national security implications, 5G technologies plan to use the following three electromagnetic spectrums: high band (also called millimeter wave, or MMW), which operates between around 24 and 300 GHz; mid band, which operates between 1 GHz and 6 GHz; and low band, which operates below 1 GHz collectively known as sub-6.

Each spectrum provides unique opportunity for the mobile network. Low-band spectrum ensures widespread coverage, an enabler for Internet of Things (IoT) services and suburban connectivity. The low-band spectrum easily propagates through walls and has longer range of coverage. The mid-band spectrum typically offers some good mix of both coverage and capacity. The high-band spectrum will ensure high speed data rates which are crucial for high traffic demand at high network speeds at shorter ranges. Thus, 5G is expected to provide wider coverage and faster data rates than 4G.

In 2019, the World Radio Communication Conference (WRC-19) revised the international treaty that governs the use of spectrum frequencies to include several millimeter wave frequency bands for use in 5G mobile services, with around 17 GHz of spectrum identified in total globally or regionally for 5G in the 26 GHz (24.25–27.5 GHz), 40 GHz (37–43.5 GHz), 50 GHz (45.5–47 GHz and 47.2–48.2 GHz), and 66 GHz (66–71 GHz) ranges (GSMA Intelligence, 2021).

5.2.1 Technical capabilities of 5G

The 60 GHz applications provide short-range information exchange for com-munication, computing, consumer electronics, and LAN extension at very high data rates. The 60 GHz band has previously been limited in use due to its difficulty of propagation through air and materials, in addition, its short range and the cost of producing products based on this frequency in commer-cial quantities prohibited its use in previous generations of mobile networks.

With the recent availability in various international markets of high power, license exempt spectrum in this range, the economies of scale in supporting commercial and scientific applications in the 60 GHz band has changed dramatically.

5.2.2 Advantages of 5G

Compared with 4G, some generic benefits of 5G include higher data transfer speed, lower latency, the data-powering backbone for edge computing, higher capacity bandwidth, and higher network availability, among others.

5G promises to break the limitations of current mobile communication network, such as bandwidth, latency and speed. These constrains or limitations restrict the kind of application, the size of data, the amount of information that could be transmitted and the number of devices that could potentially connect to the Internet and access cloud computing resources. Lifting most of the constraints of data transfer and capacity opens more opportunities for innovation and allow diverse devices and applications to be connected to the Internet via the mobile network.

This means increased productivity for organizations, convenience for domestic users, improved way of life with the influx of wearables, controls, and accessibility to almost all kinds of devices and services that are improved by the capabilities of 5G.

5.2.3 Shortcomings of 5G

5G shortcomings are technically not in the technology that powers 5G, but rather what the possibilities of 5G brings on board, what 5G will enable, empower, and transform.

5G promises vast information and knowledge in the hands of both criminals and law-abiding citizens alike. It therefore requires elevated security to be able to deal with criminals who have same set of knowledge or even faster access to knowledge and intelligence.

• 5G's main use case is Artificial Intelligence (AI), IoT, and edge computing or intelligent edge. Edge computing is computing that is done at or close to the source of the data, thus, instead of relying on the cloud for computing and intelligence, we are relying on where the data is generated so that there is no need to send data for processing in the cloud, thereby bringing the power of the cloud closer to users. This computing paradigm will decentralize computing from the cloud and increase storage and processing points whereby increasing the attack surface and requiring additional security measures by stakeholders. This will be an attractive avenue for cybercriminals since most organizations do not have the necessary security tools and expertise to protect their infrastructure.

- IoT is computing and the interconnection of everyday objects via communication networks such as the Internet, which previously had no interface to access a network and the Internet as whole. This simply means that nearly all electronic and electrical devices and appliances such as iron and washing machines will be connected to the Internet, and objects like footballs, shoes, and cars will be equipped with technologies such as Wi-Fi and Bluetooth protocols that support connection to the Internet. What this means is that there will be more devices with many vulnerabilities, as such devices will not be equipped with high-end security chips due in part to technical challenges, feasibility and constraints due to weight, size, and the type of material used for such devices.

- Operational technology (OT) devices such as gauges and sensors, which previously were operating with airgap, are now being connected to the Internet for productivity and efficiency. Empowered by 5G for efficient and faster data processing and transfer could mean consequent increase in the attack surface and exposure to cyberattacks.

- Similar to surge and proliferation of mobile phone devices, 5G will cause the demand for wearable devices, domestic sensors, lighting systems, and others to surge in connectivity. These devices, without efficient security, could put users' lives in danger, for example, heartbeat monitors could be manipulated to the detriment of the wearer, and organized criminals can sell noncompliant devices with backdoors to surveil people illegally without their knowledge.

- To fully realize the advantages of IoT coupled with 5G, billions of IoT devices will now have the needed capacity to connect to the Internet. With each of these IoT devices generating data, transfer of data to the cloud for computing and processing will be inefficient. Therefore, there will be the need to process the data on the IoT devices or closer to the IoT rather than having the data and computing done in the cloud. Consequently, there will be an increase in edge computing. These edge computing devices now becomes an attractive vector for cybercriminals to mount attacks for obvious reasons as data processing will be in the hands of users to secure and protect.

• Lifting database application and transferring it over to remote location within seconds will now be a possibility. It will be faster and quicker to steal huge amount of information remotely, as there is no limitation of transfer speed over the 5G communication channels.

• Real-time ultra-high-definition data communication and high-speed bandwidth with minimal to no latency, will also spike the use of robotic medical assistance and remote doctor assistance to conduct very complex and delicate surgical procedures, a breach in such systems could prove fatal, the possibilities of 5G must be adopted with extra care and security awareness.

• Possible catastrophe with shortage of semiconductors with the surge in IoT devices. Shortage of semiconductors is a possibility with 5G as every household device in addition to existing electrical and electronic devices globally will need some microchip of a sort for connectivity and processing. If vendors are unable to meet demand for IoT devices, a possible spike in cost of device could be the result as individuals and organizations alike would like to own IoT for the benefits it brings on board. Another possible result is the introduction of substandard devices by criminals which could have severe impact on consumers.

5.3 5G Adoption and Potential Benefits to Ghana

According to MTN Ghana's CEO, 5G should be expected somewhere around 2026 but will have to be rolled out around 2023 as COVID has accelerated digitalization (Monzon, 2021).

MTN is a leading mobile service operator in Ghana with a market share of mobile data of 68.49% as of March 2021, with an estimated population of 31,108,574, the total subscriptions of data in the country at the end of March 2021 were 22,936,004 with a penetration rate of 73.73% for the period according to the NCA (National Communication Authority, 2021).

This high percentage of mobile services penetration in part could be attributed to the surge in mobility-as-a-service providers such as Uber, Bolt, and Yango; food delivery services such as Glovo; and e-commerce sites such as Jumia, Melcom, Tonaton, Jiji, etc., which entice people with the convenience and cost-saving provided.

Almost all major banks have an online platform of a kind that enables customers to transact businesses online.

With the COVID-19 restrictions on movement, quite a number of schools in Ghana did adopt the Internet-based video streaming teaching and learning platforms.

Video conference solution provider, Zoom, recognizing the limitation of bandwidth capacity in major parts of the world including Ghana, provisioned a solution that enables Zoom to work in low-bandwidth environments.

The solution is simple traffic prioritization where audio and screen sharing are prioritized over video during bandwidth congestion to ensure that Zoom service and user experience are not degraded. Though it works, there is a compromise.

Few schools that provisioned online video and teaching platform faced Internet connection challenges such as low-bandwidth capacity, poor quality of connectivity among others. Majority of schools and individuals at home were struggling with limited Internet connectivity bandwidth, others were also having difficulties getting access at various places in the country due to lack of Internet services in those areas.

All these are indicators that the demand for high-speed data in Ghana is eminent. 5G services are considered major breakthrough to fix nearly all Internet connectivity issues that have to do with capacity, speed, range, and bandwidth.

Companies such as Samsung, Huawei, Apple, Oppo, Xiaomi, Nokia, and OnePlus have already started shipping 5G phones to Ghana. According GSMA Forum, Sub-Saharan Africa will have 332 million IoT connections by 2025, with applications across a range of use cases (GSMA Intelligence, 2019). It is evident that as mobile services and platforms grow, 5G capacity and services are needed for efficient operation of such services. As interest in data services soar and the supply of 5G mobile devices increase on the Ghanaian market, the ground is being prepared for the full take-off of 5G in the country. Most Ghanaian homes are already equipped with smart home appliances with high-speed wireless and Ethernet connection ports and interfaces which will connect over the 5G fixed wireless broadband in the immediate future. As the customer-end equipment gets into the hands of subscribers, telecom services providers need only to roll out their services. There is currently no projection of future fallen trend in mobile and Internet services adoption which means that bandwidth demand will only increase at least into the foreseeable future.

According to (Ericsson, 2021), a global telecommunications infrastructure provider, the total global mobile data traffic – excluding traffic generated by fixed wireless access (FWA) – reached 49 EB per month at the end of 2020 and is projected to grow by a factor of close to 5 to reach 237 EB per month in 2026. The fifth generation (5G) mobile technology provides the platform to enable other technologies and services such as edge computing, artificial intelligence, Internet of Things, and machine learning. These services are deployed and integrated together. 5G has the capacity to connect more devices per area than 4G. The CNN business insider states that 5G will make it possible to connect many, many more devices to the network. At a minimum, 5G networks can support **1 million devices per every square kilometer** (Thakor, Razzaque, & Khandaker, 2021). **In other words, 5G network can allow anything to be connected to the Internet without affecting performance.** As mentioned by Ericsson, the leading provider of 5G network equipment in the United States, 5G is designed to be able to connect much more types of devices than mobile smartphones. This capacity leads to the surge of IoT devices because anything can now be connected to the Internet.

Connecting virtually anything means that generating more data even from unusual sources than could be achieved previously. This Big Data is what Artificial Intelligence (AI) and Machine Learning depend on for learning and decision making. Machine Learning has a voracious appetite for data during both development and production as stipulated by Philip Russom, Ph.D., director of TDWI Research (Russom, 2018). To sum it all up, we can differentiate between 5G and 4G based on three factors, which as are speed, capacity, and latency as shared by Clare Duffy, a CNN technology Business writer (Duffy, 2020). Each of these differences is a capability of 5G that can be harnessed for greater good.

Benefits to education in Ghana: Online video teaching and learning

Educational institutions in Ghana can take advantage of 5G speeds to implement full online video teaching and learning platforms. This has already proven to be feasible during the COVID-19 outbreak and the lockdown period as cost-effective means of teaching and learning. It saves travel time, hosting fees and for some cost of transportation to and from schools. An obvious challenge faced by many is the low speed and data throughput which makes the teaching and learning experience a sub-par.

Many schools are likely to adopt a whole online teaching and learning model, as this will not only save cost for students and parents, but it will also reduce the pressure on our scarce classrooms and teaching resources the country is currently faced with. Government intervention for real-time online video teaching platform will go a long way to enhance learning and teaching while saving cost. It will also bridge the gap between well-resourced schools and under-resourced schools since all students can benefit from the same quality of teaching.

Quality teaching in most rural areas is a major challenge. 5G with its high capacity and range simply means that rural Internet connectivity is feasible and cost effective. Ghana education service and other educational institute can connect some of these schools over 5G networks to access what is accepted as quality teaching and learning from well-established and endowed schools in the urban areas without having to travel or spend a fortune as a subsidized subscription service from the government. This will provide an important platform for all students across the country. A remote student at a village will get to benefit from quality teaching in his home rather than under a tree. 5G speeds can make this a reality.

Benefits for security services

According to Vodafone, a world leader in 5G technology, 5G is expected to support up to 1 million connected devices per 1 square kilometer, compared to around 2000 with 4G, which is crucial for Internet of Things, where lots of additional devices such as drones and cars will eventually be connected to mobile networks.

Ghana could harness this capacity of 5G for the security services.

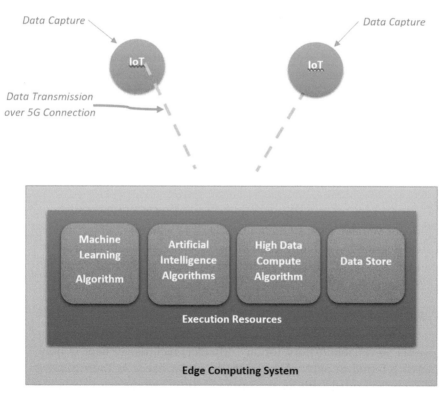

Figure 5.2   Edge computing diagram.

All police personnel and vehicles can be conspicuously equipped with smart connected body cams and smart video recorders on vehicles that will conduct high definition video surveillance in real time at any location and transmit them to a central source to provide the service with intelligence at national scales without affecting the performance and quality of the network.

According to the Superintendent of Police Mrs. Naomi Acquah, the availability and access to quality data, had been the bedrock of all success-ful law enforcement activities. Quality data was accessed by its accuracy, completeness, consistency, timeliness, validity, and uniqueness. Quality elec-tronic data ensures a better and safer place (Joy Online, 2021).

The security services in addition to the use of IoT for surveillance, could integrate it into edge computing for even better results, where these connected smart video surveillance solutions, will rely on AI and machine learning to capture vehicle registration number, raise alert, check insurance details, cap-ture vehicle color, location, and other vital information in real-time. Edge computing is one of the services to be enabled by 5G. The main benefits edge solutions provide include low latency, high bandwidth, device processing and data offload as well as trusted computing and storage (Ericsson, 2021).

Edge computing is a distributed computing architecture that brings compute resources closer to the users, data and devices that need them, improving response times and saving network bandwidth costs (Horwitz, 2021). IoT will capture the data at the source of incidence but the analysis of the data using AI and Machine Learning algorithms to produce intelli-gence will be done at the edge. Edge computing will enable users to be able to sense, capture, and analyze massive amounts of data at the location of creation and obtaining global visibility, management, and deeper analysis – or even the creation of machine-learning models (Casey, 2021). Thus, edge computing is a computationally efficient, secure, private, cost-effective way to utilize IoT at scale without running the risk of data breaches or network overloads (Grand View Research, 2021). It is without doubt that the Ghana Police Service in particular and Ghana as whole, need effective and reliable solution to enhance its data capture and processes for law enforcement which affects everyone living in Ghana.

Road safety in Ghana

Currently offensive road users are given notes written on papers, therefore, there is no reliable history for any offender; records are kept in books and can be referred to only when the books are at hand. At most places traffic rules are compiled with by road users only in the presence of police offi-cers. Edge computing could also be used to identify traffic offenders to tag, track and follow, raise alarms and flag them all in real-time. The police can capture thumbprint of road offenders which will be circulated nationwide, enabling such offenders to be identified anywhere within the country for similar offenses. By so doing road carnage will reduce since every offense is available to all police personnel in real time. The transport industry and the State transport corporations, especially long journey travelers, can adopt real-time video surveillance in their buses, cars, and trucks. Land borders can be equipped with surveillance and be effectively monitored in real time to make up for the few numbers of security patrols placed along the borders.

Elections in Ghana

One crucial and important area of Ghana's constitutional and democratic governance is election. It affects the entire country of Ghana, from the services, to health care, to chieftaincies and the clergy. Elections stretch every limited resource available, from logistics to security, human resources, consultants and experts. It is one aspect of democracy that causes fear, tension, panic, and violence across Ghana. Peace songs flood the airwaves throughout the election periods. Religious councils and traditional leaders use their platforms to educate Ghanaians about the need for peace. Nearly all service security personnel are drawn to assist in maintaining peace and order in the country. This goes without saying that elections are most crucial aspects of Ghanaian culture. The writer believes that 5G and its associated technologies could be used to solve most of the challenges of Ghanaian elections such as multiple voting, ballot snatching, multiple ballots counts, multiple registrations, rejected ballots, and underage voting which most often are the source of conflicts during the election process.

According to the Electoral Commission of Ghana, the 2020 elections were held in more than 33,000 polling stations nationwide (Electroral Commission Ghana, 2021). Voting materials must be shipped to each of these locations, with its accompanied security. The elections procedure in Ghana basically involves voter registration and voter identification against registered voter list. Some security features implemented to detect and prevent voting fraud includes the use of indelible ink, valid voter ID cards, and physical observation of voters to determine if the person is underage. The writer attempts to apply 5G capabilities and the technologies enabled by it to attempt to solve most of the challenges faced during Ghanaian elections.

The solution via 5G adoption

As quoted by Jason "In fact, we're so sensitive to being watched that even just a drawing or photograph of a pair of eyes influences our decisions"

(Goldman, 2014), this mechanism of being watched provides a solution for the above-mentioned electoral situation in Ghana, except this time it would not be a pair of physical eyes watching, but millions of eyes of surveillance devices across all polling stations, and these are described below:

- All voting systems must be connected together via 5G to central monitoring system and collation dashboards at the premises of all stakeholders.

- The voter register must be centralized and connected to all polling stations and other stakeholders over the 5G network with read only access for security.

- A read only version of polls or list of all votes casted must be transmitted from all polling stations to the EC offices and other stakeholders.

- Biometric data and facial recognition tools and applications for identification must be checked against the read only database.

- Drones can be used for monitoring throughout the polls; all elections observers could deploy smart connected drones to capture live videos from all polling stations of interest to them.

- Facial recognition tools will not be used for identification purposes only but also capture everyone that comes into the vicinity of the polling station or attempts to disrupt the process.

According to Time magazine, surveillance has become an everyday part of life in most developed societies which has been aided by an explosion in AI-powered facial-recognition technology (Campbell, 2019). This implies that these technologies are readily available and with the rollout of 5G their application will have significant impact on the Ghanaian society as a whole.

Centralized identification, facial recognition, and biometric tools will detect and block double voting at any polling station in real-time using AI and machine learning algorithms. Multiple registrations by a single person and identity theft will be detected through the use of biometric and facial recognition tools integrated into the centralized voting system.

The centralized and connected database, voting system, identification, and counting and monitoring systems will allow voters to vote anywhere there is a polling station without having to travel to respective places of registration.

The benefits of all things connected together with high-speed data, AI and machine learning when harnessed effectively can potentially reduce election cost and its associated challenges Ghana is faced with during every election year.

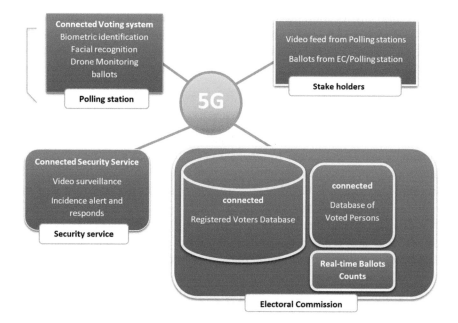

Figure 5.3   Diagram of 5G voting system and monitoring.

Ballot counting that takes days to complete, could be done in real time while voting is underway since the voting systems are connected. With the available bandwidth capacity of 5G and its range, every polling station could be monitored remotely regardless of how remote the polling station is. This, the writer believes, will enhance the voting exercise and ensure credible elections with the use of fewer resources, that will not cost Ghana a fortune and will most importantly curb the tension, fear and panic that always grip the country during elections.

Legal and regulatory challenges

Although 5G is a new technology, the rules and regulations as defined in the NCA regulatory documents regarding electronic communication will equally apply to 5G services and technologies. A major challenge that Ghana will face is the ability to monitor and track thousands if not millions of IoT devices and the huge volume of data to be generated every second. There will also be concerns of privacy and surveillance.

According to Azena, an award-winning smart video surveillance company in Germany funded by Bosch group, the growing demands for smart video surveillance and a rise in IoT will drive further integration of camera systems into any location possible. IoT brings with it video surveillance at an epic scale and that has always been a challenge for most governments as far as privacy of individuals is concerned and how that information could be used.

There have always been conflicts between security and privacy. When it comes to 5G services, a key concern will be privacy of individuals when the state begins to roll out IoT surveillance for the purpose of security, safety, peace keeping, and intelligence gathering.

Even though there has not been much talk about privacy and surveillance in Ghana as compared to other parts of the world where such facilities already exist, there shall be a public outcry about the use of such data, and the discomfort of being watched anywhere. There shall also be concerns about individual privacy and state security.

The fine line between security and privacy must be addressed with supportive legal and regulatory frameworks with regards to usage and distribution and handling of individual data. It is likely that people may not feel comfortable being watched everywhere and that will be a major challenge the government must begin to address through stakeholders. The Ghanaian public has accepted CCTV cameras at shopping malls and on our streets, knowing the purpose. Having the same infrastructure over 5G, simply means that the captured data is no longer local to the organization and that could mean a lot more to them. 5G has not enjoyed unanimous acceptance globally, there has been misinformation and disinformation about it. Any infrastructure, framework, or service to be built on 5G must begin with trust, as Frances X Frei rightly puts it (Frei & Morriss, 2020). Penalties and punishment for misuse of such information must be considered and addressed clearly before the implementation of the system.

5.4 Potential Security Threats of 5G Adoption in Ghana

- It is likely the full potential of 5G will not be known at the time of adoption and implementation. Some of the pitfalls will surface as the usage of 5G grows. Prior to that, it will be very difficult to patch every security hole that might be present during the adoption period and criminals can take advantage of such unknown vulnerabilities.

- Acquisitions of multiple cell sites towers that require more security presence could prove costly to most providers, who in an attempt to cut cost may leave certain sites unmanned which will lead to physical attacks on infrastructure and security of the network.

- Lack of regulation, standards, and compliance especially for domestic devices could allow cybercriminals to ship sensors with back doors to steal data and information. IoT cameras for homes with minimum security could be breached by cybercriminals.

- It will now be possible for drones to spy on citizens and capture and transfer high definitions data from any location in the country in real time without any hindrance, and criminals can now use autonomous vehicles to hide their identities and avoid being caught by security officials.

- Criminals will have a new attack surface in the autonomous vehicles, as vehicles can be hijacked without having to be physically present, due to the ability to remotely operate vehicles.

- IoT brings a new kind of security challenge that requires users of these everyday objects turned computing devices, to understand how to operate and update security patches. There is the likelihood that most users of these devices will be coming from the use-and-forget era, where an iron is only plugged and used to iron and nothing more. Secondly majority of users of these traditional devices are not all computer literates, and there is a tendency of people not understanding how to secure their devices, etc., which will equally lead to weaknesses in the overall infrastructure of these devices. A survey shows that 47.9% of Ghanaians own basic phones, which is higher than 46.1% that uses smart phones (National Communications Authority & Ghana Statistical Service, 2020).

- AI that requires human and machine interaction usually through voice commands, will also pose a new kind of security challenge, as there is no AI as of yet that supports our local languages, and Ghanaian locals, having different pronunciation from westerners, could possibly lead to insurgence of middle men who will be required to assist people with language and communication barriers. Most AI devices are activated by simple voice commands, when every home is equipped with such AI criminals can play voice recorders to activate AI functionalities to execute certain tasks to gain unauthorized access.

5.4.1 Recommendations for security challenges for 5g in Ghana

Below are a few recommended ways to curtail some of the above listed potential security challenges with 5G adoption in Ghana:

- Biometric identification and authentication should be key requirement in autonomous vehicles.

- Data rights and protection management that enforces restrictions on copy and transfer of files for unauthorized access must be strictly implemented.

- Implementation of device-dependent data protection systems that will render any data illegally copied onto another device unintelligible, to safeguard against illegal data transfer.

5.5 Conclusion

When 5G takes off fully, it will cause other prominent technologies like IoT to surge. IoT in turn will cause massive spike in data generation of billions per second. Such massive amount of data being generated from multiple IoT devices will not be efficient for cloud processing, and that is where edge computing shines.

Data generated by security services via IoT devices nationwide will need to be analyzed and computed; however, the current powerful compute locations are in the cloud. Pushing billions of data from millions of IoT devices in real time to the cloud for processing and computing in real time will be very challenging; therefore, edge computing takes the processing from the cloud to devices closer to the IoT which generated the data for processing which will result in faster response times. Everyone could potentially wear a bodycam that can stream high-definition videos in real time.

5G will open a whole lot of opportunities and challenges that cannot be foretold, and without proper regulation framework in place we will be at the mercy of global social media and application developers to sell products and services that can even compromise the security of the country. Just as people are using YouTube and digital airwaves to broadcast all sorts of materials at the detriment of Ghanaians, without strict regulation because it is simply possible and easy to do.

The possibility of real-time city scape services can cause people to take undue advantage of 5G to unlawfully capture and broadcast high-definition videos from locations without authorization. People will become broadcasters of high-definition video from any location of choice. Criminals will take advantage and entrepreneurs will monetize this opportunity. Without proper legal frame work, such possibilities will become uncontrollable.

References

Lewis, M. (2021). *Telstra pushing 5G into regional and rural Australian communities.* Retrieved from Mobilecorp, Sydney-Australia: https://www.

mobilecorp.com.au/blog/telstra-pushing-5g-into-regional-and-rural-australian-communities

Ahmed, A. (2021). Lightweight digital certificate management and efficacious symmetric cryptographic mechanism over industrial Internet of Things. *Sensors*, 2810.

Akinsolu, M., Sangodoyin, A., & Adeyemi, K. (2021). Design considerations and data communication architecture for national animal identification and traceability system in Nigeria. *IST-Africa 2021 Conference Proceedings, IST-Africa Institute/ IIMC.*

Ali, R., Pal, A., Kumari, S., Karuppiah, M., & Conti, M. (2017). A secure user authentication and key-agreement scheme using wireless sensor networks for agriculture monitoring. *Future Generation Computer Systems, 84*, 200 - 2015.

Aranda, J., Sacoto-Cabrera, E., Haro-Mendoza, D., & Astudillo, F. (2021). 5G networks: A review from the perspectives of architecture, business models, cybersecurity, and research developments. *Novasinergia, 4*(1), 6–41.

Arizton Research. (2022). *Latin America Data Center Market Report Scope 2021–2026.* Arizton Research.

Beavers, I. (2018). Intelligence at the edge Part 3: Edge node communication. *Technical Report, 2018. Analog Devices Engineer Zone.*

Bestsennyy, O., Gilbert, G., Harris, A., & Rost, J. (2021). *Telehealth: A quarter-trillion-dollar post-COVID-19 reality?* McKinsey Insights.

Biamis, A., & Curran, K. (n.d.). 5G security and the Internet of Things. In *Security and Organization within IoT and Smart Cities.* CRC Press: USA.

Bronson, K. (2019). Looking through a responsible innovation lens at uneven engagements with digital farming. *NJAS-Wageningen Journal of Life Sciences*, 90 - 91.

Brummer, T. (2021). Cybersecurity in beyond 5G: use cases, current approaches, trends, and challenges. *Communication Systems XIV, Technical Report IFI-2021.02 Chapter 3.*

Cabral, E., Silva Castro, W., Florentino, D., Araujo Vianna, D., daCosta Junior, J., Pires de Souza, R., et al. (2018). Response time in the emergency services. Systematic review. *Acta Cir. Bras., 33*(12), 1110–1121.

Campbell, C. (2019). *What the Chinese Surveillance State Means for the Rest of the World.* Retrieved October 19, 2021, from https://time.com/5735411/china-surveillance-privacy-issues/

Casey, K. (2021). *Edge computing and IoT: How they fit together.* Retrieved October 19, 2021, from https://enterprisersproject.com/article/2021/3/how-edge-computing-and-iot-fit-together

CEA. (2014). Cybersecurity Enhancement Act of 2014 (Enacted January 1, 2021). *Public Law*, 113–274.

Chow, W. (2021). *The global economic impact of 5G" Global Technology, Media and Telecommunications (TMT).* PWC.

Cordeiro, M. (2021, 128). *Brasil ganha primeira antena 5G voltada para o agro no sul do país.* Retrieved from Digital Policy Law: https://digitalpolicylaw. com/sercomtel-realizara-proyecto-piloto-5g-en-areas-rurales-de-brasil/

Correa Lima, G., Lira Figueiredo, F., Barbieri, A., & Seki, J. (2020). Agro 4.0: enabling agriculture digital transformation through IoT. *Revista Ciência Agronômica -*.

Cosby, A., Manning, J., Fogarty, E., & Wilson, C. I. (2021). *Assessing real time tracking technologies to integrate with identification methods and national traceability requirements.* North Sydney: Final Report, Project V.RDA.2005. CQ University Australia. Meat and Livestock Australia Ltd.

Creese, S., Dutton, W., & Esteve-González, P. (2021). The social and cultural shaping of cybersecurity capacity building: a comparative study of nations and regions. *Personal and Ubiquitous Computing, 1 -15.*

CSRIC 7. (2020). *Report on recommendations for identifying optional security features that can diminish the effectiveness of 5g security - Communications Security, Reliability and Interoperability.* USA: Working Group 3: Managing Security Risk in Emerging 5G Implementations.

Davies, J., & Goldberg, R. (1957). *A concept of Agribusiness.* Boston USA: Division of Research, Graduate School of Business Administration, Harvard University.

Dimitrievski, A., Filiposka, S., Melero, F. J., Zdravevsvki, E., Lameski, P., Pires, L. M., et al. (2021). Rural healthcare IoT architecture based on low-energy LoRa. *International Journal of Environment Research and Public Health, 18*(2021), 7660.

Dorairaju, G. (2021). *Cyber security in modern agriculture case study: IoT-based insect pest trap system.* MsC. Thesis, Jamk University of Applied Sciences, Finland.

Drougkas, A., Liveri, D., Zisi, A., & Kyranoudi, P. (2020). *Procurement guidelines for cybersecurity in hospitals.* European Union Agency for Cybersecurity (ENISA).

Duffy, C. (2020). *The big differences between 4G and 5G.* Retrieved October 19, 2021, from https://edition.cnn.com/2020/01/17/tech/5g-technical-explainer/index.html

Electroral Commission Ghana. (2021). *2020 Presidential Election Update.* Retrieved October 1, 2021, from https://ec.gov.gh/

Ericsson. (2018). *A guide to 5G network security.* Insight Report 18:000589 Uen Ericsson AB.

Ericsson. (2021). *Edge computing - a must for 5G success.* Retrieved October 1, 2021, from https://www.ericsson.com/en/edge-computing

Ericsson. (2021). *Ericsson Mobility Report - EAB-21:005137 Stockholm, Sweden June 2021.* Ericsson.

Ericsson. (2021). *Mobile Data Traffic Outlook.* Retrieved October 19, 2021, from https://www.ericsson.com/en/reports-and-papers/mobility-report/dataforecasts/mobile-traffic-forecast

ETSI. (2021). *5G security architecture and procedures for 5G System (3GPP TS 33.501 version 16.6.0 Release 16)–Technical Specification TS 133501 V16.6.0 (2021-04).* Retrieved from 5G Americas: https://www.5gamericas.org/wp-content/uploads/2021/01/InDesign-3GPP-Rel-16-17-2021.pdf

Farooq, M., Riaz, S., Abid, A., Umer, T., & Zikria, Y. (2020). Role of IoT technology in agriculture: a systematic literature review. *Electronics, 9*(2), 319.

Fontenla, E. (2016). Cooperativas agropecuarias argentinas: nueva generación de cooperativas. *Serie Documentos Nº 1; Ediciones CGCyM. Buenos Aires.*

Fonyi, S. (2020). Overview of 5G security and vulnerabilities. *International Conference on Cyber Conflict (CyCon US 2019) Defending Forward (Spring), 1*, pp. 117–134.

Forge, S., & Vu, K. (2020). Forming a 5G strategy for developing countries: a note for policy Makers. *Telecommunication Policy, 44*(7).

Frei, F. X., & Morriss, A. (2020). *Begin With Trust.* Retrieved October 10, 2021, from https://hbr.org/2020/05/begin-with-trust

Gabriel, C., & Kompany, R. (2021). *'Open RAN: ready for prime time? The operators´ perspective, Research report, April 2021.* London, UK: Analysys Mason Ltd.

Gamreklidze, E. (2014). Cyber security in developing countries, a digital divide issue. *The Journal of International Communication, 20*(2).

GCSCC. (2021). *Cybersecurity Capacity Maturity Model for Nations (CMM) - 2021 Edition.* Global Cyber Security Capacity Centre Dpt. of Computer Science, University of Oxford, UK.

Gleißner, M., Dotzler, J., Hartig, J., Aßmuth, A., Bulitta, C., & Hamm, S. (2021). IT security of cloud services and IoT devices in healthcare. *Proc. XII International Conference on Cloud Computing, GRIDs, and Virtualization (Cloud Computing 2021).*

Goedde, L., & Revellat, J. (2020). Agriculture's connected future: how technology can yield new growth. *Discussion Paper - McKinsey Global Institute, 2020.*

Goldman, J. G. (2014). *How being watched changes you – without you knowing*. Retrieved October 10, 2021, from https://www.bbc.com/future/article/20140209-being-watched-why-thats-good

Goussal, D. (2017). Rural Broadband in Developing Regions: Alternative Research Agendas for the 5G Era. In K. E. Skouby, I. Williams, & A. Gyamfi, *Handbook on ICT in Developing Countries: 5G Perspective.* Aalborg: River Publishers.

Grand View Research. (2021). *5G Infrastructure Market Size Share & Trends Analysis Report 2021–2028, Report ID: GVR-4-68038-234-1.*

Griffith, M. (2021). Balancing the Promise and the Peril of 5G: The State of Play of the United States. *5G Beyond Borders Workshop, Woodrow Wilson International Center.* USA.

GSMA Intelligence. (2019). *5G In SUb-Saharan Africa: laying the foundations.* Retrieved October 19, 2021, from https://data.gsmaintelligence.com/api-web/v2/research-file-download?id=45121572&file=2796-160719-5G-Africa.pdf

GSMA Intelligence. (2021, January). *The economics of mmWave 5G - GSMA Intelligence.* Retrieved October 12, 2021, from https://data.gsmaintelligence.com/api-web/v2/research-file-download?id=59768858&file=210121-Economics-of-mmWave.pdf

Gupta, M., Abdelsalam, M., Khorsandroo, S., & Mittal, S. (2020). Security and privacy in smart farming: challenges and opportunities. *IEEE Access, 8* (2020).

HM Government. (2020, August 18). *5G mobile technology: a guide.* Retrieved October 12, 2021, from https://assets.publishing.service.gov.uk/government/uploads/system/uploads/attachment_data/file/913179/5G_mobile_technology_a_guide.pdf

Hong, E., Ryu, J., & Lee, E. (2021). *Entering the 5g era: lessons from Korea.* Digital Development Global Practice, World Bank.

Horwitz, L. (2021). *How Industrial Edge Fuels Real-Time IoT Processes.* Retrieved October 19, 2021, from https://www.iotworldtoday.com/2021/03/18/how-industrial-edge-fuels-real-time-iot-processes/

Hurel, L. (2021). Cybersecurity in Brazil: an analysis of the national strategy. *Strategic Paper SP-54.*

IDB. (2020). Cybersecurity risks, progress, and the way forward in Latin America and the Caribbean. *Cybersecurity Report.*

IMDRF. (2020). Principles and Practices for Medical Device Cybersecurity. *International Medical Device Regulators Forum, Doc. IMDRF/CYBER WG/N60FINAL:2020.*

Immerman, G. (2021). *The Importance of Edge Computing for the IoT.* Retrieved October 10, 2021, from https://www.machinemetrics.com/blog/edge-computing-iot

ISO/IEC. (2012). *Information security techniques- Guidelines for cyber-security- ISO/IEC 27032.* Retrieved from International Organization for Standardization (ISO)/International Electrotechnical Commission (IEC): https://www.iso.org/standard/44375.htm

ITU. (2020). *Global Cybersecurity Index 2020 - Measuring commitment to cybersecurity.* Geneve: ITU-D (International Telecommunications Union).

Joy Online. (2021). *Ghana Police Service has 18,000 convicts on database – IGP.* Retrieved October 19, 2021, from https://www.myjoyonline.com/ghana-police-service-has-18000-convicts-on-database-igp/?param=

Khan, R., Kumar, P., Jayakody, D., Dushantha, N., & Liyanage, M. (2019). A survey on security and privacy of 5G technologies: potential solutions, recent advancements and future directions. *IEEE Communications Surveys & Tutorials, 22*(1), 196–248.

Koebler, J. (2017). *Why American Farmers Are Hacking Their Tractors With Ukrainian Firmware.* Vice Mother board.

Køien, G. M. (n.d.). On threats to the 5G service based architecture. *Wireless Personnal Communications, 119*, 97–116.

Kristen, E., Kloibhofer, R., Díaz, V., & Castillejo, P. (2021). Security Assessment of Agriculture IoT (AIoT) Applications. *Appl. Sci., 11*(13), 5841.

Kumar, P., & Sharma, S. (2021). An empirical evaluation of various digital signature schemes in wireless sensor network. *IETE Technical Review, 2021.*

Lieder, S., & Schröter-Schlaack, C. (2021). Smart farming technologies in arable farming: Towards a Holistic Assessment of Opportunities and Risks. *Sustainability, 13*(2021), 6783.

Lumbard, K., Ahuja, V., & Snell, M. (2020). Open Agriculture and the Right-to-Repair Community Movement. *MWAIS 2020 Proc.15th Midwest Association for Information Systems Conference, Des Moines, Iowa May 28–29.*

Maquinac. (2021). *La pulverización selectiva se afianza en Argentina.* NVS Desarrollos.

Matinmikko-Blue, M., Yrjölä, S., Ahokangas, P., & Hämmäinen, H. (2021). Analysis of 5G spectrum awarding decisions: How do different countries consider emerging local 5G networks? *23rd Biennial Conference,*

International Telecommunications Society (ITS) Sweden 21st-23rd June, 2021. Gothenburg.

MDCG. (2019). Guidance on cybersecurity for medical devices. *Medical Device Coordination Group - Doc. MDCG 2019-16, (EU) 2017/745, 2019.*

Meyerhoff, R. (2019). *Argentina gets smarter about sustainable agriculture.* Forbes.

Monzon, L. (2021). *MTN Announces 5G Roll Out Date in Ghana.* Retrieved October 19, 2021, from https://www.itnewsafrica.com/2021/09/mtn-announces-5g-roll-out-date-in-ghana/

Mueller, K., Coburn, A., Knudson, A., Lundblad, J., McBride, T., & MacKinney, C. (2021). *Characteristics and challenges of rural ambulance agencies–A brief review and policy considerations.* USA: Rural Policy Research Institute (RUPRI).

Nai-Fovino, I., Neisse, R., Hernandez-Ramos, J. L., Polemi, N., PoRuzzante, G., Figwer, M., et al. (2019). A Proposal for a European Cybersecurity Taxonomy. *EUR 29868.*

National Communication Authority. (2021, June 1). *INDUSTRY INFORMATION - TELECOM SUBSCRIPTIONS FOR MARCH 2021.* Retrieved October 19, 2021, from https://www.nca.org.gh/assets/Industry-Report-March-2021-.pdf

National Communications Authority & Ghana Statistical Service. (2020, March). *Household Survey on ICT in Ghana.* Retrieved October 12, 2021, from https://statsghana.gov.gh/gssmain/fileUpload/pressrelease/Household%20Survey%20on%20ICT%20in%20Ghana%20(Abridged)%20new%20(1).pdf

NIST. (2018). *Framework for improving critical infrastructure cybersecurity, Version 1.1.* USA: National Institute of Standards and Technology.

Nowak, T., Sepczuk, M., Kotulski, Z., Niewolski, W. A., Artych, R., Bocianiak, K., et al. (2020). Verticals in 5G MEC-Use Cases and Security Challenges. *IEEE Access, 9*(2020).

NUKIB. (2018). The Prague proposals-chairman statement on cybersecurity of communication networks in a globally digitalized world. *European Union 5G Security Conference, Prague 3 May 2019- National Cyber and Information Security Agency (NÚKIB).* Czech R.

Racovita, M. (2021). *Industry briefing: cybersecurity for the Internet of Things and Artificial Intelligence in the AgriTech sector.* London, UK: Industry Briefing PETRAS National Centre of Excellence for IoT Systems Cybersecurity.

Roddy, M., Truong, T., Walsh, P., Bado, M., Wu, Y., Healy, M., et al. (2019). 5G Network Slicing for Mission-critical use cases. *2019 IEEE 2nd 5G World Forum (5GWF),* (pp. 409–414).

Rose, D., Wheeler, C., Winter, M., Lobley, M., & Chivers, C. (2021). Agriculture 4.0: Making it work for people, production, and the planet. *Land Use Policy, 100.*

Rugeles, J., Guillen, E., & Cardoso, L. (2021). A technical review of wireless security for the Internet of Things: Software Defined Radio perspective. *Journal of King Saud University - Computer and Information Sciences.*

Russom, P. (2018). *Data Requirements for Machine Learning.* Retrieved October 10, 2021, from https://tdwi.org/articles/2018/09/14/adv-all-data-requirements-for-machine-learning.aspx

Samsung. (2019). 5G Launches in Korea: get a taste of the future. *White Paper.*

Scaramuzza, F. (2021). Adelantos tecnológicos y el futuro inmediato de la agricultura de precisión. *(Conference) INTA (Instituto Nacional de Tecnología Agropecuaria. Manfredi, Córdoba. Biblioteca Fundación CIDETER, Argentina 2021.*

Shimabukuro, I. (2021, 12 5). *Brazilian government inaugurates its first 5G antenna in a rural area of the country.* Retrieved from Olhar Digital: Brazil

Świątkowska, J. (2020). Tackling cybercrime to unleash developing countries'digital potential pathways for prosperity. *Commission Background Paper Series, 33.*

Talukder, A., & Hass, R. (2021). AIoT: AI meets IoT and web in smart healthcare. *WebSci'21 Companion, June 21–25, 2021, Virtual Event, UK.*

Thakor, V., Razzaque, M., & Khandaker, M. (2021). Lightweight cryptography algorithms for resource-constrained IoT devices: A review, comparison and research opportunities. *IEEE Access, 9*(2021).

Thales. (2021). *5G technology and networks (speed, use cases, rollout).* Retrieved October 10, 2021, from https://www.thalesgroup.com/en/markets/digital-identity-and-security/mobile/inspired/5G

Tomasin, S., Centenaro, M., Seco-Granados, G., Roth, S., & Sezgin, A. (2021). Location-privacy leakage and integrated solutions for 5G cellular networks and beyond. *Sensors, 21*(2021), 5176.

Trakadas, P., Sarakis, L., Giannopoulos, A., Spantideas, S., Capsalis, N., Gkonis, P., et al. (2021). A cost-efficient 5G non-public network architectural approach: key concepts and enablers, building blocks and potential use cases. *Sensors, 21*(2021), 5578.

Tursunov, R., Lenox, J., & Cleave, J. (2019). *Digital healthcare South Korea. Market intelligence report 2019.* UK: Department for International Trade (DIT) - Intralink Group.

Unit 42. (2020). *2020 Unit 42 IoT Threat Report.* CA USA: Palo Alto Networks, Santa Clara.

USTDA. (2020). *ICT project opportunities in Argentina, Brazil and Paraguay. A resource guide for US industry.* USA: U.S. Trade and Development Agency.

Wazid, M., Kumar, A., Shetty, S., Gope, S., & Rodrigues, J. (2021). Security in 5G-enabled Internet of Things communication: issues, challenges and future research roadmap. *IEEE Access, 9*(2021).

Weissman, C. (1969). Security control in the ADEPT-50 time-sharing system. *Proc. 35th AFIPS Conference*, (pp. 119–133). USA.

Winkler, K., Fuchs, R., Rounsevell, M., & Herold, M. (2021). Global land use changes are four times greater than previously estimated. *Nature Communications, 12*(2501).

WTI. (2020). *Annual Report 2020.* World Tele Health Initiative, Santa Barbara, CA. USA.

Wu, T., Yang, L., Lee, Z., Chu, S., & Kumar, S. (2021). A provably secure three-factor authentication protocol for Wireless Sensor Networks. *Wireless Communications and Mobile Computing.*

Yazdinejad, A., Zolfaghari, B., Azmoodeh, A., Dehghantanha, A., Karimipour, H., Green, A. G., et al. (2021). A review on security of smart farming and precision agriculture: security aspects, attacks, threats and countermeasures. *Applied Sciences, 11*(16), 7518.

Zhai, Y., Xu, X., Chen, B., Lu, H., Wang, Y., & Li, S. S. (2020). 5G Network-enabled smart ambulance: architecture, application, and evaluation. *IEEE Network.*

Zhang, A., Heath, R., McRobert, K., & Llewellyn, R. (2021). Who will benefit from big data? Farmers' perspective on willingness to share farm data. *Journal of Rural Studies.*

6

Mapping the Iranian Policy Network for Protecting Users' Data on Platforms

Somayeh Labafi[1], Ali Darvishi[2] and Hadi Moghadamzadeh[3]

[1]Iranian Research Institute for Information Science and Technology (IranDoc), Iran.
[2]Allameh Tabataba'i University, Iran.
[3]University of Tehran, Iran.
Email: Labafi@irandoc.ac.ir; darvishi@gmail.com; Moghaddam.h@ut.ac.ir

Abstract

Emerging concerns about user data security on platforms have forced governments to intervene in this field. Since no comprehensive policy document regarding the protection of user data has been yet developed in Iran, research efforts to identify a network of policymakers and barriers in this regard are required. This research aims to employ the institutional mapping approach to identify and map governmental and private actors of the policymaking network for protecting the personal data of Iranian users on platforms. This research has mapped a network of actors in this field in order to identify the network of data protection policymakers in Iran. After identifying these actors and the connections between them, the gaps and the huge number of actors as well as outputs of the policymaking network for user data protection will be identified. The results also show that the inefficiency of the policymaking network in the development and execution of policies is mainly due to the conflicts of interest of its main actors.

6.1 Introduction

Advances in emerging technologies have caused new issues and challenges for the policymaking and regulatory systems of countries. Moreover, the rapid pace of growth and dynamism in this field has placed traditional policymaking

131

systems in a dilemma of innovation strengthening or adherence to traditional rules and policies (Evens & Donders, 2018; Nooren et al, 2018). As an emerging technology, media platforms today provide new opportunities for large-scale conversation and interaction for their users (Parker et al, 2016: 46). Although these new capacities and opportunities provide numerous benefits, they have their own disadvantages. Such platforms collect the personal data of their countless users, classify them into different demographic, geographical, psychological, and behavioral categories, and finally employ them for various purposes. In fact, the personal and public data of users can be easily collected from their behaviors on these platforms as well as their digital footprint (Simonofski et al, 2021; Matheus, Janssen, & Maheshwari 2018). To gain insight into the big data they access, such platforms are increasingly seeking to use big data solutions to advance their business, political, etc., goals (Gorwa, Binns & Katzenbach, 2020).

Growing concerns about user data security have forced governments to intervene in this field. Personal data protection is referred to by different terms in different countries, such as information privacy, data privacy, and data protection, all of which are a basis for developing legal frameworks to protect the personal data of users (Jayakar, 2018; Reini, 2019). Considering the serious political, cultural, and economic consequences of user data analysis and processing, governments must establish regulatory and legal frameworks, such as the EU's General Data Protection Regulation (GDPR), the Brazilian Lei Geral de Proteção de Dados (LGPD), and the South African Protection of Personal Information Act (POPIA), in this field. Such frameworks should protect personal data to prevent intrusion and manipulation of users' choices based on illegal data processing, prevent interference in democratic processes, and protect free dialogs (Labafi, 2020). Considering the above-mentioned requirements, Iran as a developing country also needs such policies and rules to protect user data because its citizens have warmly welcomed media platforms in recent years. There is a need for a set of principles and norms to be developed with practical and effective applications to guide behaviors of platform's ecosystem regarding the user data. Although Iran's laws and regulations occasionally and sporadically protect user data in cyberspace, the shortcomings and inadequacy of these rules and regulations made Iran's Ministry of Information and Communications Technology develop a bill called "Protection of Personal Data" in 2019 in order to protect user data in cyberspace. However, this bill has not yet been enacted by the Islamic parliament of Iran. Evidence suggests that the main barriers to the enactment and enforcement of this bill are the lack of integration between policymaking institutions on social medial platforms, unclear boundaries of such ecosystems, absence of a trustee or a main actor in this field, the rapid pace of technology growth, the changing desires of users, and so on (Labafi et al., 2020). Considering the rapid growth rate of media platforms as well as the data collection and analysis by these platforms in Iran, inattention to these issues and lack of appropriate policies can lead to the poor monitoring of the way these platforms collect and analyze user data and also users' dissatisfaction with and distrust in these platforms.

Since no comprehensive policy document regarding the protection of user data has been yet developed in Iran, research efforts to identify a network of policymakers and barriers in this regard can be work. It is necessary to map a network of actors in this field in order to identify the network of data protection policymakers in Iran. After identifying these actors and the connections between them in this network, it will be possible to understand the gaps and overloaded regarding the actors as well as outputs of policymaking network for user data protection. This study hence aims to employ the institutional mapping approach to identify and map governmental and private actors of the policymaking network for protecting the personal data of Iranian users on platforms.

6.2 Literature Review

6.2.1 A review of media platforms regulatory

The emergence of big data and its application in various sectors reveals the significance of data regulation (Bauer, M. et al, 2014). Much of the big data is the product of user interactions on digital platforms. Platforms act as data gatekeepers. These platforms provide cost-effective services to users, but, in turn, control and influence their behavior (Nooren et al. 2018). Digital platforms have affected all aspects of social life, such as markets, social behaviors, and political processes. Contrary to initial optimism, disinformation, fake news, and hate speech on digital platforms that distorts social discourses are common these days (Perscheid, Ostern & Moormann, 2020). Frequent breaches of privacy and the abuse of security holes are examples of disadvantages of such platforms. Therefore, policymakers have started regulating data on these platforms to protect public interests, including privacy and user data (Napoli, 2019). It is noteworthy that policymakers should intervene in the regulation of digital platforms at the right time and place in order to cause positive results for public interests (Labafi, Tokbaeva, Jalalpour, 2020).

The effects of digital platforms on the public interest justify the government intervention in this field. Public interest primarily justifies the government intervention in digital platforms from an economic perspective. Accordingly, the prevention of market failure is usually considered the main reason for the governmental intervention (Evans and Gawer, 2016). External influences, market power, information asymmetry, and economies of scale are some reasons for this market failure (Lund, 2016; Van Gorp and Price, 2017). Even in the case of a market failure, the cost-effectiveness of governmental interventions (including the cost of policy failure) should be compared to the situation the government does not intervene in this field. On a larger scale, government intervention can be justified in a normative framework based on cultural diversity, rights and freedoms and so on. Either positively or negatively, digital platforms can also affect the public normative interests such as freedom, privacy, and information protection (Sozur, 2018). Another reason for the government intervention in digital platforms is related to concerns that are referred to as "Generativity". This term refers to the extent to which a platform can be a breeding ground for other phenomena or become something beyond the original intent of its developer. Considering such capabilities of digital platforms, policymaking in this field is a necessary but, yet, complex and ambiguous process.

The term "platform governance" has been raised to reduce the complexity and ambiguity of policymaking in this field. "Platform governance" has rooted on the "media governance", which was first defined by Puppis (2010) as a set of principles, rules, and law aimed at regulating media systems. After identifying the main actors of the media regulation, he proposed a pattern in which the concept of governance replaces the concept of "centralized regulation". As long as the concept of media governance is discussed, we address some approaches to media policy such as self-regulation, co-regulation, and encouragement of stakeholder participation. In fact, the concept of media governance has more to do with the structure of policies rather than their content. Freedman (2010) also defines "media governance" as all formal or informal, national or transnational, and centralized or decentralized mechanisms aimed at organizing media systems. Media governance is a framework that can both restrict or motivate the media systems (Puppis, 2010). After the emergence of large platforms and the need to regulate them, the concept of "platform governance" was also derived from "media governance". Three major actors of governance are government, companies, and nongovernmental organizations (NGOs). All these actors are partially engaged in any policymaking action. This is the network of actors presented by Abbott and Snidal (2009). The competencies of each of the actors are needed for different phases of policymaking, including agenda setting, monitoring, and execution.

A review of the platform governance literature shows that this concept is defined based on five essential elements (Perscheid et al. 2020). These five essential elements are decision right (who has the authority and responsibility for making decisions about platforms?), transparency (documentation that allows users to understand how the platform is governed or, in other words, what its information policy is?), accessibility (how much the platform's ecosystem is accessible?), trust (the relationship between the governor, users, and the platform supplements), and incentivization (incentives that the platform gives to internal or external developers to improve and secure its operations). Any applications of the governance approach for the regulation of digital platforms should take into account all of the above-mentioned elements.

6.3 Background of Data Protection Policies

We need to review the first legal documents on user data protection to find more about its background. The first data protection law was passed in 1950 and enforced in 1953 in the European Union (Goethem, 2018: 7). Privacy and protection of personal data received greater attention in the mid-1960s and 1970s. Following the rapid advances in information technology and growing concerns about privacy, the German federal government enacted the world's first national data protection law in the early 1970s (Bitar and Bjorn, 2017: 9). Similar laws related to the protection of personal data were also passed in Canada in 1977, France, Austria, Denmark, and Norway in 1978, and Luxembourg in 1979. Later, almost all European countries began to regulate the protection of personal data (Serzhanova, 2012: 4). The European Data Protection Directive was also adopted in 1995. A major step taken in this regard was the adoption of the General Data Protection Regulation (GDPR) by the European Union in 2016; all EU member states were required to follow these regulations from 2018. Because of its innovative and accurate features and supports, GDPR is the most comprehensive legal framework for data protection. Updating the principles set out in the 1995 EU Data Protection Directive to guarantee the right to protect personal data, GDPR emphasizes strengthening public rights, strengthening the EU internal market, facilitating the international transfer of personal data, and setting global data protection standards (Philipp Jan, 2016).

After scandals such as Cambridge Analytica and the like, policymakers faced a challenge which was referred to as "The Darker Narrative of Platform Capitalism" by Pascal (2016). The public opinion calls for the regulation and control of platforms; some examples are requests for the separation of Facebook from Instagram and WhatsApp as well as the need for platforms to be accountable for user-generated content. The European Commission began investigations about issues of digital platforms in 2015 and proposed the basics of policy on digital platforms to be a roadmap for developing policies in this field (Testa and Marelli, 2018). The drafts of the Digital Service Act (DSA) and Digital Market Act (DMA) are examples of regulatory efforts in this field. In addition to the European Commission, other countries have also developed policies in an attempt to regulate the relationship between digital platforms and national interests (Gorwa, 2019). Due to the lack of a comprehensive law on information privacy in a country such as the United States, personal data are partially protected by other legal frameworks, for example, the 2018 California Consumer Privacy Act (CCPA). This act contains comprehensive provisions for consumer privacy that are similar to some GDPR requirements (Korpisaari, 2019). CCPA requires many companies to change their business processes to comply with a new set of consumer privacy rights.

Following the popularity of digital platforms in Iran and the migration of traditional media audiences to these platforms, there were concerns about the economic, political, and social effects of such platforms. As a result, policymakers intervened in this field to protect user data. Although some policies have been developed (Ambrose & Ausloos, 2013; Russell, 2017), no specific document has been formulated in Iran for data protection. Some Iranian legal provisions, including the Law on Computer Crimes adopted in 2009 and the Law on Free Dissemination and Access to Information adopted in 2009, partially protect personal data in some cases. Some provisions of these laws protect personal data in some cases. For instance, as contained in Article 17 of the Law on Computer Crimes, anyone who publishes one's private or family voices and videos or other secrets without their consent, in a way that harms them or tarnishes their reputation, shall be sentenced to imprisonment from 91 days to two years or a fine of five to forty million rials, or both. Since it is necessary to obtain one's consent before getting access to or publishing one's private information and personal data, the disallowed access and dissemination of one's personal information is considered a crime.

Apart from such minor and incomplete supports, the draft of "Protection of Personal Data Bill", published on the website of Iran's Ministry of Information and Communications Technology in June 2018, is first the document that aims to take effective steps toward protecting personal data. However, a review of this draft reveals that it lacks the necessary transparency and accuracy and, as a result, it would be difficult to enforce. For example, Paragraph A of Article 2 of this draft defines personal data as any data that, alone or in combination with other data, directly or indirectly identifies the owner of the data by referring to an identifier. This definition seems inaccurate when compared to the definition presented in GDPR, because this paragraph does not specify whether personal data are merely exclusive to natural persons or include legal persons as well. It is noteworthy that the laws on the protection of personal data in different countries are clearly subject to natural persons. Absolutely, the conciseness and ambiguity of a legal document allow for various interpretations and, as a result, differences in its enforcement. The policymaking network is one of the causes of developing late and ambiguous policies regarding personal data protection in Iran. One solution that can reduce the complexity and facilitate the formulation of more efficient policies is to map a policymaking network and then define the relationship between institutions in this network as well as their interests. This study hence aims to find an answer to the following questions:

- How is the institutional mapping of data protection policymaking in Iran?

- What are the existing documents concerning data protection policymaking?

- Who are the main actors in the data protection policymaking network?

- How is the data protection policymaking network?

6.4 Methodology

This study aimed to map the network of personal data protection policymaking with institutional mapping approach. The authors attempted to collect and analyze the relevant data in each phase to identify existing policy documents, the main actors in this network, and the relationship between them. Figure 6.1 shows the procedure of institutional mapping.

First of all, laws that were somewhat relevant to data protection were identified and reviewed. Since no specific law on data protection has yet been passed in Iran and there are merely some provisions used for this purpose as the case may be, all the laws whose provisions have been used in this regard over the years were reviewed in this study. These laws include the Law on Computer Crimes adopted in 2009 and the Law on Free Dissemination and Access to Information adopted in 2009, which partially protect personal data in some cases. Independent policy documents developed by the government or other legislative bodies in recent years were also reviewed. It is noteworthy that none of these policy documents has been enacted and enforced.

Table 6.1 provides a summary of the reviewed laws and policy documents.

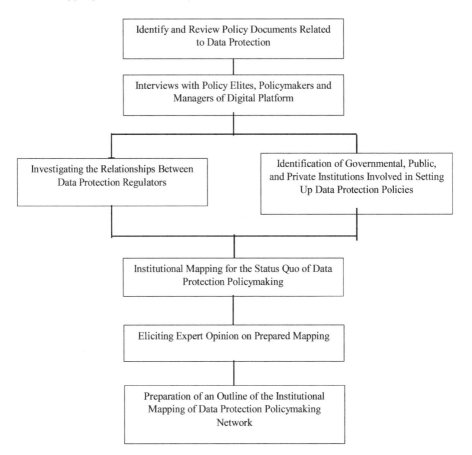

Figure 6.1   Institutional mapping for the data protection policy network.

The review of these documents reveals that policymakers have felt the absence of an independent law on the protection of user data and tried to develop special documents to protect user data over the past three years. Nevertheless, none of these documents has been yet approved and enforced because the main actors of the policymaking network cannot reach an agreement on their common interests. This is mainly because the developers or proposers of these draft documents often either do not properly recognize or ignore the interests of other institutions. For example, the majority of platforms and NGOs oppose the "Cyberspace Users Protection Plan", which was proposed by the Islamic parliament of Iran in 2021, because they argue that this plan takes into account the interests of neither media platforms nor users. Recently presented by the Research Center of the Islamic parliament of Iran, this plan has some disadvantages and

Table 6.1 Iranian laws and policy documents related to personal data protection.

Number	Title	Year of development or enactment	Proposer(s)/developer(s)	Status
1	Law on Computer Crimes	2009	Islamic Parliament of Iran and Ministry of Information and Communications Technology	Enacted
2	Law on Free Dissemination and Access to Information	2009	Islamic Parliament of Iran and Ministry of Information and Communications Technology	Enacted
3	Islamic Penal Code	2013	Government and Islamic Parliament of Iran	Enacted
4	Requirements of the national information network bill	2017	Supreme Council of Cyberspace	Enacted
5	Personal Data Protection Bill	2018	Ministry of Information and Communications Technology	Non-enacted
6	Cyberspace and Digital Economy Bill	2020	Islamic Parliament of Iran	Non-enacted
7	National Information Network Architecture Master Plan	2020	Supreme Council of Cyberspace	Enacted
8	Plan to support the development and competitiveness of platforms providing basic and applied services of the National Information Network	2020	Islamic Parliament of Iran	Non-enacted
9	National Data and Information Integration Plan	2020	Islamic Parliament of Iran	Non-enacted
10	Data and Information Dissemination Plan	2020	Islamic Parliament of Iran	Non-enacted
11	Cyberspace Users Protection Plan	2021	Research Center of the Islamic Parliament of Iran	Non-enacted

drawbacks that may prevent it from protecting personal user data in cyber-space. For example, the scope of this plan is vague and unclear, it does not explicitly support user data against all types of access, the regulatory structure of this plan will not be efficient due to the strong conflict of interest, it adds to the complexity of filtering, and this plan will make it more difficult for domestic platforms to operate. Moreover, the negative and criminal attitudes to digital platforms and lack of incentives for platforms regarding the protection of user data are among other shortcomings and problems of this proposed legal document.

The most comprehensive legal document in this regard is probably the "Personal Data Protection Bill", which was developed by Ministry of Information and Communications Technology in 2018. However, the draft of this bill has its own problems when compared to similar international documents such as GDPR. One of these problems is that it considers no proper executive guarantee in the case of personal data violation. For instance, punishment in the criminal liability section of this draft (Articles 65 through 68) is based on the relevant provisions of the Islamic Penal Code. Although such a reference is not objectionable in nature, it will be effective when the financial penalties contained in Article 28 of the Islamic Penal Code can cause deterrent effects. However, the deterrence of financial penalties contained in the Islamic Penal Code is itself a matter of debate. Therefore, similar to Article 83 of GDPR, financial penalties should be commensurate with the inflation rate to cause greater deterrent effects. In addition, financial penalties can be also determined proportionally to the inflicted damage to better compensate the damage and loss of individuals. It can be hence concluded that even the most comprehensive Iranian document on personal data protection is subject to shortcomings that can lead to the ineffectiveness of any legal document.

After reviewing the legal documents related to personal user data protection, in-depth and semi-structured interviews with experts were conducted. The participants consisted of policymakers, managers of platforms, and experts in cyberspace who were selected based on their willingness to participate in the study and their relevance to the research subject. Table 6.2 provides a list of participants and the reasons why they were selected. Some of the questions asked in interviews were as follows: "How is the policy-making process regarding the protection of user data?", "How is this process formed?", "Who are the main actors of this process?", What are the implications of user data protection policymaking process?", "What are the challenges of draft laws prepared so far?", and "Why has no single policy document been developed and enacted in this regard?".

Table 6.2 List of participants.

Number	Position	Reasons for selection	Area of expertise
1	Legal advisor to the National Information Technology Organization	Expertise in data privacy	Regulation of platforms
2	A researcher on platform regulatory	History of researching domestic and international documents related to digital platforms	Policymaking and regulation
3	A researcher on platform regulatory	Expertise in data governance	Data governance
4	Manager of platform	High level of conflict with policymakers	Management of a platform
5	Manager of platform	Direct involvement in data policy issues	Management of a platform
6	Manager of platform	Direct involvement in data policy issues	Management of a platform
7	Manager of platform	Direct involvement in data policy issues	Management of a platform
8	A policymaker of digital platforms	Preparation of legal drafts in the field of user data protection	Policymaking on digital platforms
9	A policymaker of digital platforms	Preparation of legal drafts in the field of user data protection	Policymaking on digital platforms
10	A policymaker of digital platforms	Preparation of legal drafts in the field of user data protection	Policymaking on digital platforms
11	A data scientist	Expertise in big data analysis	Big data analyzer
12	A data scientist	Expertise in big data analysis	Big data analyzer

After coding and analyzing the content of interviews, eight main themes were extracted that explained why no single legal document has been developed and enacted on user data protection and challenges of draft laws on data protection prepared by each of the policymaking bodies (Table 6.3).

The interviews with experts revealed that structural and content issues are among the main reasons why no single policy document has been yet developed, enacted, and executed to protect user data on digital platforms. Some examples of content issues are "Low expertise of actors in data

Table 6.3   Codes and themes extracted from the interviews.

Number	Theme
1	Conflict of interests of main actors in the policymaking network for user data protection
2	Nontransparent interests of public and private actors regarding the type and extent of applying user data
3	Parallel efforts of actors for preparing legal drafts
4	The absence of a learning cycle in the policymaking network of user data protection
5	Nonclarity of policymakers in determining the type of data to be protected
6	Inefficiency of the existing policymaking structure for user data protection
7	The lack of positive attitudes to encourage platforms to execute relevant documents
8	Infeasibility of all provisions of the prepared drafts

protection policymaking network", "disagreements over interests of actors in this policymaking network", and "the lack of positive attitudes to encourage platforms to execute relevant documents". Moreover, "inefficiency of the existing policymaking network for user data protection", "infeasibility of the prepared drafts", and "parallel efforts of actors for preparing drafts" are some examples of structural problems.

In the next stage, the authors tried to identify the main policymakers in this field by monitoring various organizations and institutions involved in this field as well as reviewing the information obtained during the study. Based on the results, the institutions involved in policymaking about user data protection were divided into three groups: sovereign institutions, governmental institutions, and specialized associations (especially belonging to the private sector). Tables 6.4, 6.5, and 6.6 present a list of these institutions and organizations.

After identifying the main institutions and organizations involved in this field, the next stage was to determine the relationships between them. Figure 6.2 presents a map of policymaking institutions and relationships between them. This map is based on the categorization of policymaking actors for user data protection, that is, sovereign institutions, governmental institutions, and private associations. As Puppis et al. (2019) state, policymaking actors may be involved in different levels such as initial drafting of policies, regulation of industry activities, and execution of policies. Therefore, based on the pyramid designed by Puppis et al. (2019) regarding different levels of policymaking, the involvement level of the main actors in

Table 6.4 Sovereign institutions involved in the policy network for user data protection.

Number	Sovereign institutions	Is it directly related to setting data protection policies?
1	High Council of Cyberspace (National Cyberspace Center, High Commission for Regulation, High Commission for Content Promotion, High Commission for Security)	Yes
2	Islamic parliament of Iran (Research Center, Cultural Commission, National Security Commission, Judiciary and Legal Commission)	Yes
3	Expediency Discernment Council (Legal and Judicial Commission, Defense and Security Commission, Scientific, Cultural and Social Commission)	No
4	Islamic Republic of Iran Broadcasting (Iranian Audiovisual Regulatory Body)	Yes
5	Supreme Council of the Cultural Revolution	No
6	Islamic Development Organization	No

Table 6.5 Governmental and executive organizations involved in the policy network for user data protection.

Number	Governmental and executive organizations	Is it directly related to setting data protection policies?
1	Ministry of Industry, Mines, and Trade (e-Commerce Development Center)	No
2	Ministry of Culture and Islamic Guidance (Center for Development of Information Technology and Digital Media, Center for Development of Culture and Art in Cyberspace)	No
3	Ministry of Information and Communications Technology (Information Technology Organization)	Yes
4	Ministry of Interior (E-Government Development Center)	Yes

user data protection in this map was defined in three areas of policy, regulation, and execution.

After preparing the initial mapping, it was sent to participants in research to review the relationship once again and propose any necessary change. Finally, the status quo of policymaking about user data protection in Iran can be outlined as shown in Figure 6.3.

Table 6.6   Private and public sectors involved in the policy network for user data protection.

Number	Companies and private groups that are most often consulted about setting up data protection policies	Reasons for the presence in the policy network
1	Internet Service Providers (ISPs)	ISPs have the executive power to implement policies
2	Managers of major VOD platforms	Access to data of a large number of users
3	Managers of major IP TV platforms	Access to data of a large number of users
4	Iranian ICT Guild Organization	Some members of the specialized commissions of this organization are experts in Internet services
5	National E-Business Association	The representatives of platform-based e-businesses are present in this association

Figure 6.2   Levels of policymaking and main actors in user data protection.

6.5 Discussion and Conclusion

A prerequisite for a platform's ecosystem is the regulating and execution of policies that both facilitate the activities of the participants in this ecosystem (users, platforms, etc.) and cause the development of this ecosystem. Since platform's ecosystems are a newly-emerged phenomenon, there is still much debate about the policies of such ecosystems and many countries are still attempting to develop appropriate policies for this field. The novelty

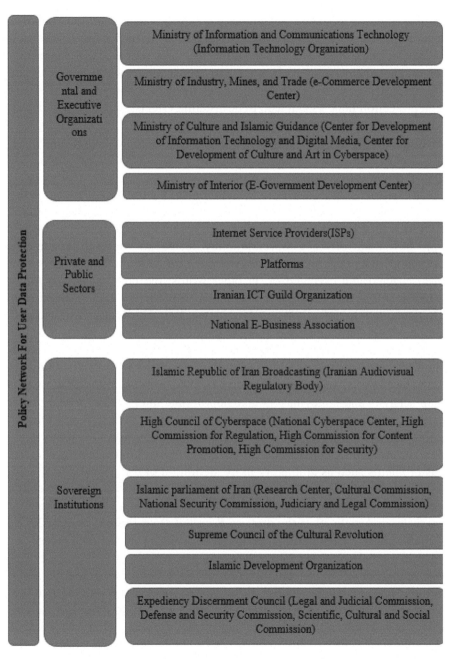

Figure 6.3 Actors in Data Protection Policy Network.

of platform's ecosystems necessitates the development of various policies, including those related to user data protection. Data ownership is now considered a major issue related to platform's ecosystem, because large platforms take advantage of their users' data to engage in social, political, and any other type of interventions. That is why the lack or poor execution of user data protection policies can lead to devastating effects, such as users' dissatisfaction with and distrust in platforms. Therefore, different countries have developed and sometimes enacted various policy documents in recent years. Nevertheless, there are still concerns about policies for user data protection on digital platforms.

As a developing country with an immature but growing platform's ecosystem, Iran also needs to develop policies to protect user data on digital platforms. In recent years, some governmental and nongovernmental institutions have made efforts to develop policy documents within the framework of a policymaking network. However, none of these policy documents have been enacted and executed or, in other words, the policymaking network established for user data protection failed to produce satisfactory outputs. The inefficiency of this policymaking network in the development and execution of policies is mainly due to the conflict of interest of its main actors. Identification and categorization of actors to this network, review of relationships within this network, and evaluation of the outputs through institutional mapping can serve as a beacon of activists and experts in this field. It can also help activists and policymakers take appropriate interventions based on the existing shortcomings and problems. This study hence attempted to identify different components of the policymaking network for user data protection.

Analysis of policy documents showed that the draft of these documents has problems such as ambiguity in the definition of user private data, inappropriate scope, the infeasibility of some of their provisions due to divergent executive processes, and inattention to the interests of all stakeholders. For example, in Article 39 of the draft of *Personal Data Protection Bill* (2018), the authorities and institutions appointed to monitor personal data processing have no expertise in data protection and seem to be less efficient in the execution of this document. Another problem of this document is its unclear scope of authority; for instance, when a platform is controlled jointly by two persons, one of whom is Iran, or when the controller is a foreigner and the processor is an Iranian, it is not clear whether they are subject to this document.

When it comes to the structure and stakeholders of the policymaking network for user data protection, the results showed that stakeholders in none of the three levels (sovereign, governmental, and private) have reached the necessary convergence and agreement to develop and execute policies in this area. Issues such as conflict of interests of different stakeholders, multiplicity of proposed documents, and parallel executive work add to the complexity of this situation. It can be hence concluded that the existing policymaking network for user data protection in Iran lacks the convergence and expertise necessary for developing policy documents that all stakeholders agree on and strive to enact and execute them.

Based on the study findings, some recommendations can be made to change the status quo and reduce the problems in this area. First of all, it is necessary to appoint a specialized institution, which has already been assigned similar duties by the law, to draft a comprehensive document for user data protection. To take into account executive requirements, this institution should be an executive one such as the National Information Technology Organization. This can add to the specialized quality of draft documents. Moreover, the "authority" and "interests" of policymaking institutions are recommended to be considered in drafting such a document in order to achieve the convergence and consensus necessary for its enactment and execution. Finally, policymakers are recommended to take advantage of the experience of large platforms to implement internal policies of user data protection and also to develop more appropriate and enforceable policies.

References

Ambrose, M. L., & Ausloos, J. (2013). The right to be forgotten across the pond. *Journal of Information Policy*, 3. 1–23.

Bauer, M. et al. (2014), '*The Costs of Data Localization: Friendly Fire on Economic Recovery*', ECIPE, Occasional Paper No. 3/2014, European Centre for International Political Economy, Brussels, Belgium.

Bitar, H. and J., & Bjorn. (2017). *GDPR: Securing Personal Data in Compliance with new EU-Regulations A7009N GDPR: Securing Personal Data in Compliance with new EU- Regulations.* Luleå University of Technology (pp. 1–71).

EC. European Commission. (2019). Data Protection Certification Mechanisms Study on Articles 42 and 43 of the Regulation (EU) 2016/679 (pp. 1–255).

EU Court of Justice. (2018). Uber France SAS v Nabil Bensalem. Judgment of the Court (Grand Chamber) (Case C-320/16). https://eur-lex.europa. eu/legal-content/EN/TXT/?uri=CELEX: 62016CJ0320.

Evans, D.S., and A. Gawer. (2016). "The Rise of the Platform Enterprise: A Global Survey." The Emerging Platform Economy Series, Vol. 1. The Center for Global Enterprise. https://www.thecge.net/app/ uploads/2016/01/PDF-WEB-Platform-Survey_01_12.pdf.

Evens, T. and Donders, K. (2018), *Platform Power and Policy in Transforming Television Markets*. London.

Freedman, D. (2008), *The politics of media policy*, Cambridge, UK: Polity Press.

Goethem, A. van. (2018). The Effects of Brexit on GDPR Implementation An investigation into data protection legislation within the United Kingdom (pp. 1–62)

Gorwa, R. (2019). The platform governance triangle: Conceptualizing the informal regulation of online content. *Internet Policy Review*, 8(2).

Gorwa, R., Binns, R., & Katzenbach, C. (2020). Algorithmic content moderation: Technical and political challenges in the automation of platform governance. *Big Data & Society*, 7(1)

Jan Philipp, A. (2016). How the GDPR Will Change the World. In European Data Protection Law Review (Vol. 2, Issue 3, pp. 287–289).

Jayakar, K. (2018), 'Media policy', in A.B. Albarran, B.I. Mierzejewska, and J. Jaemin (eds.), *Handbook of Media Management and Economics,* 2nd ed., New York and London: Routledge, pp. 178- 200.

Korpisaari, P. (2019). Finland: A Brief Overview of the GDPR Implementation (pp. 232–237)

Labafi, S. (2020). Iranian data protection policy in social media; an actor-network theory approach. In I. Williams (Ed), *Contemporary applications of actor network theory* (pp. 121- 139). Singapore: Palgrave Macmillan. Retrieved from: https://www.palgrave.com/gp/book/9789811570650

Labafi, S., kia, A. and Maleki, M. (2020). Engaging User in Academic Social Network: Identifying Effective Factors (Case study: Research Gate). *Iranian Journal of Information Processing and Management*, 36(1), pp. 33–62. Retrieved from: https://jipm.irandoc.ac.ir/article-1-4402-en.html

Labafi, S., Tokbaeva, D., Jalalpour, M. (2020). Media Innovation and its Influence on Policy-making in the Social Media Sector. Nordic Journal of Media Management, 1(4), 557–573. DOI: 10.5278/ njmm.2597-0445.6514

Lund, A. (2016), 'A stakeholder approach to media governance', in G.F. Lowe and C. Brown (eds.), *Managing Media Firms and Industries,* Berlin: Springer, pp. 103–120.

Matheus, R., Janssen, M., & Maheshwari, D. (2018). Data science empowering the public: Data driven dashboards for transparent and accountable decision-making in smart cities. *Government Information Quarterly*. 35(3)

Napoli, P. N. (2019). User Data as Public Resource: Implications for Social Media Regulation, *Policy and Internet*. doi: 10.1002/poi3.216.

Nooren, P., van Gorp, N., van Eijk, N., & Fathaigh, R. (2018). "Should We Regulate Digital Platforms? A New Framework for Evaluating Policy Options" '*Policy and internet*' Directorate-General for Internal Policies of the Union 'Volume10' Issue3 Pages 264–301.

Parker, G.G. et al. (2016), '*Platform Revolution: How Networked Markets Are Transforming the Economy - and How to Make Them Work for You*', New York: W. W. Norton & Company.

Perscheid, G., Ostern, N. K., & Moormann, J. (2020). Determining Platform Governance: Framework for Classifying Governance Types. In ICIS.

Puppis, M. (2010). Media Governance: A New Concept for the Analysis of Media Policy and Regulation. communication, culture and critique, 3 (2), 34–149.

Reini, P. (2019). GDPR implementation Case: Head power Oy [University of Transport andCommunications].https://www.theseus.fi/bitstream/handle/10024/166514/Reini_k7696_thesis_versio4.1.pdf?sequence=2

Russell, J. (2017), 'Google wins' right to be forgotten' battle in Japan', *TechCrunch*, https://techcrunch.com/2017/02/01/google-japan-negative-comments/ Accessed 1 February 2020.

Serzhanova, V. (2012). Personal Data Protection in the European Union under the Treaty of Lisbon. 1–7.

Simonofski, A., Fink, J., & Burnay, C. 2021. Supporting policy-making with social media and e-participation platforms data: A policy analytics framework. *Government Information Quarterly*, 38(3), 112–130.

Suzor N. (2018). Digital Constitutionalism: Using the Rule of Law to Evaluate the Legitimacy of Governance by Platforms. Social Media + Society, 8(3). 78- 94.

Van Til, H., N. Van Gorp, & K. Price. 2017. "Big Data and Competition." Ecorys Study for the Dutch Ministry of Economic Affairs, Ecorys, Rotterdam. https://www.rijksoverheid.nl/binaries/ rijksoverheid/documenten/rapporten/2017/06/13/big-data-and-competition/big-data-and-competition.pdf.

7

Empirical Investigation of the Drivers of 5G and the Mediating Role of Users' Attitudes to Achieving Word of Mouth and Willingness: A Case Study of the Microfinance Institutions of Zambia

Robert Mwiinga[1] and Manish Dadhich[2]

[1]Sir Padampat Singhania University, India
[2]Sir Padampat Singhania University, India
Email: robert.mwiinga@spsu.ac.in; manish.dadhich@spsu.ac.in

Abstract

5G could be the key to realizing industry's and society's digital ambitions. Zambia is on the verge of a digital revolution, fueled by rising broadband, Internet penetration, and the volume of data consumption. The paper examined how the following concepts/dimensions namely, "privacy risk", "subjective criteria", "speed", and "ubiquity", influence attitudes toward 5G. The theoretical tool used for this evaluation is the technology acceptance model. The cases evaluated are the microfinance institutions of Zambia. Smart-PLS was utilized to examine the causal associations between the identified concepts/dimensions as well as the mediating role of "user attitudes" toward 5G. The sample size of respondents that provided the relevant insight in the evaluation process was 480. Privacy had a strong negative impact on users toward their proclivity to use 5G. Whereas speed (data rates) had a positive impact on users toward their proclivity to use 5G. There were positive correlations between attitudes toward 5G and ubiquity, subjective norms, Word of Mouth, and readiness to pay. The associations between privacy, speed, ubiquity, subjective criteria, and Word of Mouth were mediated by attitudes regarding 5G. This study's findings could help policymakers, 5G service providers, and vendors to understand the factors that drive 5G in an emerging economy like Zambia.

7.1 Introduction

New pioneering services and business models need fair assistance to play a vital part in the digital revolution (L. Cheng et al., 2021). A contemporary hyper-connected society as the fifth generation (5G) mobile networks approach has proved its significance in the various walks of society. According to (Parcu et al., 2021), telecom infrastructures evolve to integrate telecommunication systems and computing sources to utilize state-of-the-art technologies for mobiles, delivering ubiquity and new unique service opportunities. The assimilation of vertical activities in the 5G architecture is critical for penetrating society and daily lives. As a result, industrial stakeholders ranging from service developers to cloud infrastructure providers should emphasize the potential usage of 5G (Ding et al., 2021). The 5G ecosystem should start with native 5G network capabilities, such as network sharing and edge computing. 5G-ready applications are made up of cloud-native "microservices" that run on their execution environments and deploy numerous locations. This approach uses the programmable system and the computing foundation to achieve elevated levels of scalability and agility (Jeon et al., 2020).

5G is a wireless technology that can provide citizens with better and more innovative cellular services. Different countries and areas worldwide are already deploying 5G wireless networks. Operators can make this deployment a reality thanks to two enablers. The availability of spectrum is the first enabling. Network operators now have access to the frequency bands used for 5G-related signals that have already taken place in certain nations and will eventually take place in all regions. The availability of 5G new radio equipment is the second enabling (Jorquera Valero et al., 2022). The application of 5G is enormous in every industry. The technology is poised to become immensely powerful once it is completely implemented, augmenting economic growth. 5G's speed, volume, less power consumption, and quick communication response time enable a slew of new goods and services. As a result, the economic and societal benefits could be vast (Rossi, 2021).

On the contrary, few studies (Dadhich, Purohit et al., 2021; Flaherty et al., 2022) suggested that the development of 5G networks may raise privacy difficulties, owing to the innovative interfaces, common environment, and other players, all of them have different privacy viewpoints and interests. 5G networks, AI–ML–DL understandings, and the IoT are causing security and secrecy problems that necessitate regulation to help various entities achieve their privacy goals. Administrative law is the most incredible way to promote 5G privacy goals. However, because the launch of 5G has augmented mobile charges, customer approval of the technology remains shaky (Shah et al., 2021).

In the words of (Oughton et al., 2021), the willingness of consumers to allow this high tariff is a worry for 5G telecoms operators. Even though 5G is fast emerging and the issues above are crucial for telecoms businesses, researchers have rarely highlighted these issues. Therefore, this study employed an improved form of the technology acceptance model (TAM) that included numerous manifests linked to 5G usage and evaluated the impact of various 5G characteristics on customer attitude. For instance, pervasive 5G positions boost communication response times and more precise user location determination, resulting in higher-quality services. Telecommunications providers should consider consumer worries about data security and 5G issues when advertising 5G in the future. 5G's rapid reaction pace and the pervasive effects of high-density platforms were all variables included in the TAM (Davis, F. D.; Bagozzi, R. P.; Warshaw, 1989). Furthermore, the influence of subjective constructs of colleagues and family are frequently considered part of the TAM. Some users are particularly concerned about 5G's privacy issues, while others are more interested in the technology's advantages. Regulatory attention is a personal element that influences whether more attention is put on the benefits or drawbacks of 5G. Users looking for a bargain are interested in the benefits of 5G. The users with a preventative mindset, on the other hand, are more concerned about the drawbacks of 5G (Hoeschele et al., 2021). As a result, their regulatory emphasis orientation influences customer attitudes regarding 5G. Thus, the regulatory attention of consumers was also explored in this study.

The 5G era in Africa officially began in 2020, when two of the continent's largest mobile service providers, Vodacom and MTN, launched commercial 5G mobile and fixed wireless access (FWA) services in South Africa. Indeed, after the COVID-19 pandemic, the African government awarded temporary spectrum in the 3.5 GHz range; the move came sooner than predicted. Telma Madagascar and Cable & Wireless Seychelles have declared plans to establish commercial services in Africa. In contrast, 5G testing have been undertaken in Egypt, Gabon, Kenya, Morocco, Nigeria, Uganda, and Zambia (Manvinder Singh Pahwa, Manish Dadhich, Jaskaran Singh Saini, 2021; Siwale, 2021). In Zambia, telecommunications businesses can also undertake 5G advertising campaigns. MTN, Zambia's top mobile service provider, launched its fifth-generation experimental mobile (5G) network in

2021, becoming the country's first mobile service provider for high-speed Internet. MTN has officially opened the path for Zambia's new mobile network era in collaboration with the Chinese ICT giant and leading 5G vendor Huawei, making it a pioneer of Zambia's evolving mobile service provider (Kpodar & Fund, 2014).

Modern microfinance is still relatively new in Zambia, and financial services are restricted and poor even by regional standards. Zambia has also trailed other East African countries in establishing a regulatory framework for microfinance organizations. In a country with 17.5 million people and an estimated adult population of 8.1 million, outreach remains inadequate compared to the potential market. According to a 2019 FinScope data poll, 59.30% (4.8 million) of Zambian individuals are financially included. In contrast, the Association of Microfinance Institutions of Zambia (AMIZ) and other sources assess that the industry's outreach is around 3,00,000 people, compared to a potential need of over two million. According to the Microfinance Information Exchange (MIX), Zambia's microfinance sector had 71,978 borrowers and a gross loan portfolio of $6.5 million as of December 2015. However, because not all MFIs report to the MIX Market, this figure may be exaggerated (Dadhich, Hiran et al., 2021; Rajapakshe, 2021; Singh, 2021).

The current study complements the body of knowledge on 5G adoption in two ways. First, the work analyzes the impact of subjective norms, ubiquity, speed, and privacy on users' attitudes to 5G. It explores whether these characteristics change consumer WoT and WoM of the selected users. Second, this study used behavior intention as a moderator to achieve WoM and WoT. This research is also organized as follows: The second section examines the literature. The conceptual model and hypotheses are presented in Section 3. The study's data and methodology are described in Section 4. The data analysis and findings are displayed in Section 5. Section 6 summarizes the discussion, conclusion, limitations, and future directions.

7.2　Literature Review and Variables

7.2.1　Technology acceptance model

The TAM, a behavioral intention model based on Davis' concept of reasoned action in 1986, was created to assess users' behavior about their acceptance of new technology. The TAM aims to find an appropriate interactive frame for understanding user behavior associated with new information system acceptance while also studying the various elements influencing this recognition. The structure offers a theoretical framework for analyzing the impact

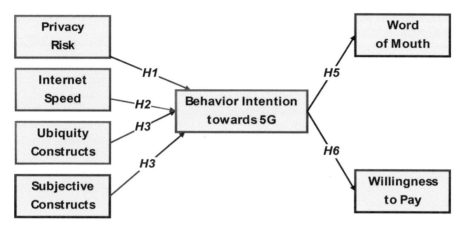

Figure 7.1   Proposed conceptual framework.

of external influences on consumers' beliefs, attitudes, intentions, and technology usage. The TAM can explain or anticipate the elements that influence information technology in any situation. According to (Davis, F. D.; Bagozzi, R. P.; Warshaw, 1989), perceived utility and easier-to-use influence users' attitudes about technology. Their behavioral intention influences people's use of information technology. Perceived usefulness and perceived ease-of-use are utilized in the TAM to explain and infer user attitudes and behavioral intentions (see Figure 7.1). As a result, the following critical characteristics linked to consumers' perceptions of 5G's usefulness and conveniences were included in this study (see Table 7.1).

7.2.2 Privacy risk

A technology's privacy risk relates to the possibility that it will infringe on personal rights (Rogalski, 2021). Privacy implies liberty from any intervention and protecting a user's personal information in the online world. In general, extending privacy to the Internet is a legally recognized person right. Persons' sensitive information, such as facts, photographs, detrimental opinions, may not be illegally infringed, sought, replicated, exposed. Moreover, personal images, passwords, documents should not be leaked while surfing 5G. Privacy assessments are conducted in service-based organizations to ascertain possible risks to the company's functions related to collecting individually specific data. There should be proper legal and regulatory policy to comply with privacy norms (Kumar Naresh, Dadhich Manish, 2014).

Table 7.1   Why 5G for emerging nations (Zambia). Sources: Oughton et al., 2021; Hoeschele et al., 2021

Use cases	Advantages	Why 5G
Predictive advanced analytics	– reduce downtime – reduce maintenance cost – reduce equipment replacement frequency	– enhance device density – Augment reliability – Costeffective
Precise supervising and control	– reduce faults – enhance yield – augment output	– augment the industrial output – less human interface – high productivity
Augmented reality	– decrease spend and time on preservation and repair – increase the efficiency of devices	– low bandwidth for data transformation – Ultralow latency
Manufacturing and services	– highprecision production – time consumption, and human development	– no interruption – maximize output – service facilitators
Health–medical sector	– low diagnostic cost – realtime and accuracy – augment life expectancy	– heal lifethreatening ailments – application of nanotechnology – Precision medications & IoT
Finance & economics	– predictive modeling – better governance & business policy – increase capitalization	– cope with developed 5G nation – better financial services – economic growth and development

According to (Rana et al., 2021), improving user perceptions of privacy and security could improve usage intention and customer satisfaction.

H₁: Customer attitude toward 5G is related to the perceived privacy risk of 5G.

7.2.3 Speed

According to (Hui et al., 2020), the most prevalent uses of mobile networks are observing videos, browsing music, doing e-commerce, and performing virtual games. The mobile network should provide a fast download speed and a steady connection. Once completely implemented, 5G technology will be far more powerful than any preceding technology, leading to a mobile technology transition unlike any other. 5G's pace, quantity, and much-decreased energy intake and transmission response period would enable a slew of new goods and services. As a result, the economic and societal benefits could be enormous. In the words of (Ji et al., 2020), five external elements influence Internet usage: browsing speed, exact function & computation, compatibility, security, and ease-of-use. Usually, video archives offered on video-sharing platforms are frequently enormous; they typically demand a long download period. As a result, users would be happier if video files could be downloaded more quickly. Thus, the users are concerned about telecommunication quality and speed. The essential aspect influencing customer adoption of 5G is speed. As a result, the below hypothesis is put forth:

H_2: *Consumer attitudes about 5G are more favorable and significant with a perceived speed of 5G.*

7.2.4 Ubiquity

It refers to the concept of everywhere and at any moment. Ubiquity in the context of mobile technology refers to the ability to use mobile devices anywhere, at any time (L. K. Cheng et al., 2021). (L. Cheng et al., 2021) specified ubiquity as the extent to which the ISPs allow an individual to interact and connect with others. It also enables consumers to retrieve data from the Internet without delay and location constraints. The structure allows individuals to connect to the Internet from any device, at any time and from anywhere. Users can use mobile network services anytime in a mobile business environment. Ubiquitous computing technology aims to build a new culture by combining computers, the Internet, place, and people. (Chang, 2012) was the first to examine the characteristics of a ubiquitous processing system in his research. Innovative cell phone services demonstrate how the technology brings accessibility, efficacy, utility, and ease-of-use. Furthermore, according to the findings of (Sharma & Dadhich, 2014; Tang et al., 2021), ubiquity enhances perceived usefulness. The TAM claims that a higher degree of perceived usefulness leads to a better consumer attitude toward 5G; consequently, perceived ubiquity is certainly connected with the attitude of 5G users.

H_3: *Consumer attitudes toward 5G are significant with higher the perceived ubiquity of 5G.*

7.2.5 Subjective constructs

Individuals are exposed to subjective norms when adopting or not approving habits. It considers whether the persons or classes who affect a person's decision-making agree with the action of other individuals. As a result, before engaging in certain conduct, an entity subjectively assesses whether decisive individuals in their life will approve of it (Parcu et al., 2021). According to (Dadhich, Purohit et al., 2021; Sreelakshmi Krishnamoorthy, Amit Dua, 2021), subjective customs are social pressure mechanisms. These are classified as informational and normative: (a) informational influence appears when a person supposes that a class is a trustworthy source of communication and hence adopts information from the group as facts; and (b) normative influence appears when communities form a social rule. Individuals follow the communities' beliefs and alter their perception and behavior. Various research has found that customers' subjective norms impact willingness to use new technology (Jorquera Valero et al., 2022). The following hypothesis was posited:

H_4: *Subjective norms are subsumed with customers' opinions toward 5G.*

7.2.6 Word of mouth (WoM)

According to (L. K. Cheng et al., 2021), WoM is the practice of consumers fleeting on their product or service reviews to other consumers after they have used them. Various research has indicated that optimistic WoM is conducive to selecting goods and services for consumers (Ding et al., 2021). Consumers' willingness to disseminate WoM is determined by assessing WoM posts, social initiations, and the necessary resources for sharing. As a result, customers have different motives for sharing their views. According to previous studies (Rossi, 2021), favorable WoM is generated when customers have a positive attitude toward commodities or services. (Flaherty et al., 2022) found that opinions toward an online forum increased electronic WoM intent. The study has shown that customers who have a good attitude regarding 5G technology are ready to disseminate positive 5G-WoM.

(Davis, F. D.; Bagozzi, R. P.; Warshaw, 1989) were the first to use attitude as a construct in TAM. Behavioral factors such as perceptions, attitudes, and beliefs regarding Internet banking qualities play a crucial role in adopting or rejecting technology (Shalendra Singh Rao, n.d.). Behavior intention becomes more prevalent in 5G Internet banking adoption because IBS directly includes monetary transactions, so fundamental consumer perceptions about risk and safety impact the customers' attitude. Thus, it is seen as a social function. People's attitudes are influenced by reinforcing or opposing them through contacts and mutual experiences, making it contagious (manish dadhich & naresh Kumar, 2015).

H₅: Consumers' behavior mediates in proposing Word of Mouth for 5G.

7.2.7 Willingness to pay (WTP)

Purchase intention was defined as customers' strategies to buy goods after thoroughly considering them. It refers to what customers do after forming an opinion about a commodity or service (Shah et al., 2021). Price is a critical aspect in determining whether a consumer will purchase. Assessing consumers' willingness to pay is another way to look at their behavioral willingness. The highest cost an individual is ready to pay for a particular product is known as a willingness to pay (L. K. Cheng et al., 2021). Consumers' WTP also refers to the extreme amount they are willing to pay for existing services and use leisure amenities. According to many studies (Oughton et al., 2021; Rogalski, 2021), consumers' attitudes and buying intentions influence their willingness to pay. According to previous research, consumer attitudes influence their intentions to utilize Internet banking services. Similarly, numerous studies that used attitude as a mediating variable found that attitude moderates perceived utility and PEOU on customers' willingness to use Internet services. The Table 7.2 delineated the chosen constructs and indicative literature to justify the validity of the items.

H₆: Consumers' behavior intention plays a mediating role in proposing willingness to pay for 5G.

7.3 Research Framework

This study investigates how users' attitudes regarding 5G were influenced by privacy, speed, ubiquity, and subjective norms. The researchers also explored the role of behavior intention as moderating to get WoM and willingness toward 5G. Figure 7.2 depicts the research framework for this study.

An online questionnaire was designed and delivered to Zambian respondents via social media platforms. After reading an introduction, respondents were asked to rate subjective norms, ubiquity, privacy, Internet speed, attitude toward 5G, WoM, and readiness to pay. After removing invalid questions, such as incomplete or contained incorrect responses, we got only 480 legitimate questionnaires. Table 7.3 outlines the outcomes of descriptive statistical evaluations and the variables' means and standard deviations.

7.3.1 Measuring variables

In terms of questionnaire design, the components in this study were measured using a 5-point Likert scale. The scales were adapted from prior research when

Table 7.2   Constructs and indicative literature.

Constructs	Manifests	Factor loading
Privacy	PR_1: My data is shared with other firms through 5G.	0.658
Risk (Rogalski,	PR_2: 5G keeps track of my routine habits/purchases.	0.797
2021)	PR_3: Companies are contacting me without my permission	0.828
	because of 5G.	0.708
	PR_4: 5G, in my opinion, creates specific privacy problems.	
Internet	IS_1: I get a timely reply to my response on 5G.	0.820
Speed (Ji et al.,	IS_2: Usually, the input is received rapidly on 5G.	0.864
2020)	IS_3: I can get information about 5G immediately.	0.623
	IS_4: Speed accomplishes routine tasks easily.	0.814
Ubiquity	UC_1: I'm free to use the 5G at any moment.	0.791
Constructs (L.	UC_2: I can utilize 5G in any location.	0.878
Cheng et al., 2021)	UC_3: When 5G is required, I can use it.	0.678
	UC_4: 5G facilitate my routine work promptly.	0.752
Subjective	SBC_1: My family believes that 5G is beneficial to me.	0.824
Constructs (Parcu	SBC_2: Most individuals advise me to utilize 5G.	0.879
et al., 2021)	SBC_3: The folks suggest I should test 5G.	0.853
Word	WoS_1: I'm willing to tell others about 5G.	0.526
of Mouth (Ding	WoS_2: I usually discuss the merits of 5G with others.	0.810
et al., 2021)	WoS_3: I recommend 5G to my family and friends.	0.819
Willingness	WTP_1: 5G is something I'm willing to pay.	0.876
to Pay (L. K. Cheng	WTP_2: I am willing to pay to continue to use 5G.	0.886
et al., 2021)	WTP_3: The upgrading to 5G is something I'm willing to pay.	0.820
Behavior Intention	BTG_1: I would be enthusiastic about the adoption of 5G.	0.526
toward 5G (Dadhich,	BTG_2: The idea of implementing 5G appeals to me.	0.799
Hiran et al., 2021;	BTG_3: Adopting 5G would be a fantastic idea.	0.821
Rajapakshe, 2021;	BTG_4: Benefit of 5G is the quality of life.	0.203
Singh, 2021)		

the questionnaire was converted from English to regional. Transformation and back-translation techniques were employed to ensure good cross-cultural adaption of the manifests (Dadhich, Rao et al., 2021).

7.3.2 Questionnaire design

Data on technology use can be acquired in various ways, depending on the research purpose. Different strategies increase data validity to obtain high-quality data (Hair et al., 2018). The researchers changed each item to fit our research environment using methods from previous studies. The cross-sectional assessment approach, in which questionnaires are sent to one or more users, is the most common method employed in this research. The first version of the questionnaire was penned using literature as a guide.

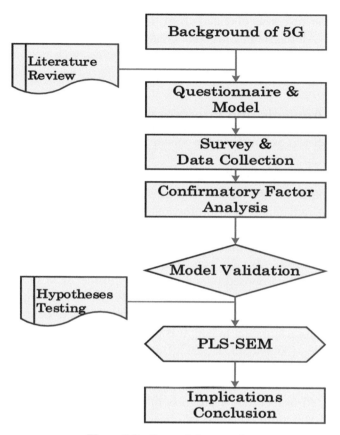

Figure 7.2 Research framework.

Table 7.3 Demographic profile.

Variables	Classification	Freq.	%
Gender	Male	255	53.10
	Female	225	46.90
		480	100.00
Education	Graduate	215	44.90
	Post-graduate	185	38.60
	Professional	80	16.50
		480	100.00
Income Level	Less than $500	285	59.40
	Above $500	195	40.60
		480	100.00
Region	Lusaka	175	36.50
	Copperbelt	125	26.40
	Central Zambia	180	37.10
		480	100.00

Based on the response obtained from the scholars, further modifications were put up. All the statements were created in English; however, they were converted into local languages depending on the research goal and subject characteristics.

7.4 Data Analysis and Interpretation

A Smart-PLS was executed on the conceptual framework to group the manifests into common constructs.

The above table delineates that the sample comprised 53.10% male and 46.90% female, 44.90% of the respondents were graduates, 38.60% were PG, and the remaining were professional 16.50%. Nearly 59.40% of the respondents had an income level of fewer than $500, and 40.60% were in the class of above $500. Similarly, 36.50% of microfinance respondents were from Lusaka region, 26.40% were from Copperbelt, and Central Zambians were 37.10%. The proposed model is analyzed using SEM in this work. A structural equation is a set of statistical models that describe how the items are related. SEM permits complex variable interactions to be expressed and provides a complete model image. It is popular in marketing and behavioral science studies (Nicolas et al., 2020) and information technology studies (Rakesh Kumar Birda & Manish Dadhich, 2019).

PLS is the preferred technique for causal prediction assessment and is better suited to theoretical progress than testing. Small samples and the complicated structural equation can benefit from PLS (Hair Jr, Joseph F., G. Tomas M. Hult, Christian Ringle, 2016). As a result, PLS-SEM is used in this work to examine the model's whole structure. The PLS frame is separated into two stages for evaluation: first, assessing the model's reliability and effectiveness; second, assessing the path framework and determining the model's potential to evaluate the structural model. The findings of these two phases are discussed in the following sections using Smart-PLS.

7.4.1 PLS measuring results

The PLS algorithm analyses the structural model to determine the importance of the path and the research model's predicting capabilities. The bootstrap procedure is then used to determine the path coefficients' significance level (Hair, J. F., Ringle, C. M., Sarstedt, 2011). To begin, the significance of the structural equation and path coefficient is carefully assessed using confidence intervals, T statistics, and standard errors. The study's hypotheses are highlighted in Table 7.7, showing the latent manifests route coefficient and the

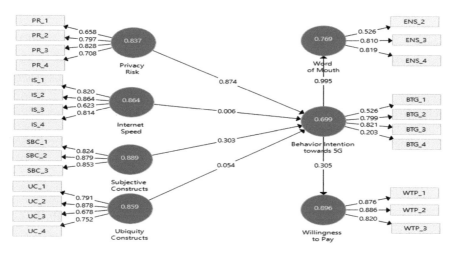

Figure 7.3 SEM-5G adoption model.

critical bootstrap ratio. The T statistic measures the estimate's consistency; a 95% confidence scale of 1.96 or above is regarded as satisfactory. Therefore, one of our research assumptions was confirmed. The consequences of each path will be explained in the following section. The model was further evaluated and validated on Smart-PLS, as shown in Figure 7.3. The discussion in the next section is based on these findings.

7.4.2 Model for measurement

(Hair Jr, Joseph F., G. Tomas M. Hult, Christian Ringle, 2016) developed a two-stage analytical approach used by the researchers. EFA and CFA analysis was used to construct validity (CV), and then structural equation modeling (SEM) was used to investigate the soil hypotheses. Similarly, Cronbach's alpha was also employed to determine the reliability of the manifests. A reliability study of all factors was undertaken in addition to the CFA. Reliability refers to whether the same results can be obtained if an investigation is repeated and whether the participants would answer in the same way until their state changes (Hair Jr, Joseph F., G. Tomas M. Hult, Christian Ringle, 2016). Within the context of reliability analysis, Cronbach's alpha statistics were used to determine the internal consistency of the factors (see Table 7.4).

All constructs have internal consistency (Cronbach's) greater than 0.759, indicating adequate reliability. The standardized factor loadings' estimated parameters exceeded the threshold value of 0.500. The average variance extracted (AVE) was also higher than the 0.50 criteria, and composite reliability

Table 7.4 Reliability and convergent validity.

Scale	Code	Chr.α	rho_A	CR	AVE	MSV
Privacy risk	PR	0.770	0.746	0.568	0.424	0.435
Internet speed	IS	0.812	0.807	0.564	0.560	0.576
Ubiquity constructs	UC	0.766	0.705	0.766	0.684	0.562
Subjective constructs	SBC	0.875	0.701	0.822	0.594	0.595
Word of mouth	WoM	0.825	0.791	0.756	0.584	0.686
Willingness to pay	WTP	0.759	0.804	0.823	0.760	0.465
Behavior intention	BTG	0.841	0.881	0.707	0.581	0.505

Table 7.5 Fornell–Larcker measure.

Constructs	PR	IS	UC	SBC	WoM	WTP	BTG
Privacy risk	0.741						
Internet speed	0.789	0.521					
Ubiquity constructs	0.781	0.408	0.615				
Subjective constructs	0.507	0.489	0.520	0.644			
Word of mouth	0.607	0.413	0.529	0.633	0.794		
Willingness to pay	0.601	0.321	0.503	0.591	0.503	0.632	
Behavior intention	0.611	0.402	0.655	0.501	0.508	0.502	0.515

Table 7.6 Heterotrait–Monotrait ratio (HTMT).

Particular	PR	IS	UC	SBC	WoM	WTP	BTG
Privacy risk	0.689						
Internet speed	0.582	0.544					
Ubiquity constructs	0.733	0.603	0.726				
Subjective constructs	0.586	0.599	0.519	0.415			
Word of mouth	0.522	0.447	0.737	0.462	0.590		
Willingness to pay	0.624	0.392	0.531	0.632	0.630	0.663	
Behavior intention	0.607	0.547	0.630	0.492	0.790	0.569	0.725

values were greater than the 0.70 (Hair et al., 2010; N. Kumar & Dadhich, 2014). Thus, the measurement simulation adopted in this study demonstrated good convergent validity (see Table 7.5). If each construct's AVE square root values are greater than the correlation coefficients between the underlying components in the research frame, discriminant validity is established.

The measurement model developed in this investigation exhibited appropriate discriminant validity and no multicollinearity issues. Similarly, HTMT denotes the mean value of item correlation through factors comparative to the geometric mean of the standard correlation for the manifests determining the same factor (see Table 7.6). In such a way, the HTMT above 0.8 indicates DV's absence. Thus, the model exhibits reasonable fit indices (Ajzen & Driver, 1992).

Table 7.7 Path coefficient and hypotheses testing.

SN	Structural path	Original sample	Sample mean	95% Conf. Inter. (LB, UB)	T- Stat	P. Val	F²
H₁	Privacy risk → Behavior intention toward 5G	0.874	0.868	(0.275, 0.247)	17.40	0.087	0.263
H₂	Internet speed → Behavior intention toward 5G	0.006	0.009	(0.321, 0.356)	3.23	0.010	0.322
H₃	Ubiquity constructs → Behavior intention toward 5G	0.054	0.041	(0.257, 0.336)	8.23	0.002	0.544
H₄	Subjective constructs → Behavior intention toward 5G	0.303	0.281	(0.208, 0.425)	7.54	0.001	0.339
H₅	Behavior intention toward 5G → Word of mouth	0.995	0.993	(0.211, 0.315)	8.45	0.020	0.291
H₆	Behavior intention toward 5G → Willingness to pay	0.305	0.314	(0.855, 0.722)	3.15	0.001	0.316

Note: $*p < 0.05$, LB-lower boundary, UB-upper boundary.

7.4.3 Validation of hypotheses

The path coefficients between constructs and the magnitude of described variation in the endogenous construct (R-square) by its exogenous construct were calculated. Table 7.7 outlines the route coefficients between constructs and robust 95% confidence ranges for their estimation. The dataset generated 5000 samples with replacement as part of a bootstrapping re-sampling technique to establish confidence intervals (Hair et al., 2018).

According to (Hair et al., 2018), SEM is a regression model problem-solving technique that addresses measurement error and attenuation correction. The researcher used PLS with maximum likelihood to test all six hypotheses under this idea.

It is examined that privacy risk positively stimulates behavior intention toward 5G (t-17.40, p-0.08, F²-0.263). Hence, the null hypothesis is excluded and reliable from the previous studies (Jeon et al., 2020; Parcu et al., 2021). Previous work also enunciated that sharing personal information concerns using 5G. Further, 5G can also keep track of my habits, including online purchases and searches. Many times, companies may also contact the person without prior permission. Thus, it may create specific privacy problems for respondents.

Similarly, speed has a positive and notable association with the intention to use 5G (t-03.23, p-0.010, F^2-0.322). Thus, H_2 was supported, and it can be inferred that higher speed better the working of these financial institutions. The customers may get a timely response using 5G, and Internet speed accomplishes routine tasks easily. These conclusions validate previous studies (Chan & Lee, 2021; Flaherty et al., 2022).

The third hypothesis demonstrated that users' attitude toward 5G adoption is influenced by ubiquity (t–8.23, p–0.002, F^2–0.544). It is an essential positive forerunner of the adoption of 5G. The respondents are free to use the 5G at any moment and promptly facilitate the routine work; this outcome is aligned with the result (Ahokangas et al., 2021).

The fourth hypothesis indicates that the attitude of 5G is positively associated with users' subjective constructs (t–07.54 p–0.001, F^2–0.339). The respondents from the microfinance field believe that 5G is conducive to sustainable development, and colleagues also encourage adding on with this new technology. The findings of this study are supported by (L. Cheng et al., 2021).

The fifth hypothesis also enunciates a positive relationship and confirms the mediating role of attitude toward 5G to achieve WoM (t–0845, p–0.02, F^2–0.291). The respondent also accepts to discuss the merits of 5G in surrounding and would like to recommend family and friends. WoM plays a substantial role, especially in marketing and technology acceptance, and the results of this work are also aligned with (Naresh Kumar, 2016; Shah et al., 2021).

According to the last hypothesis, users' attitude toward Willingness to Pay (WoP) for 5G (t–03.15, p–0.001, F2–0.31) is noteworthy. The customers of microfinance may continue to use 5G because it would enhance the work capacity, timely completion of the task, and augment productivity.

This research investigated the elements that influence the adoption of 5G technology in microfinance institutions to achieve sustainable development in Zambia. The findings support the usefulness of the TAM frameworks in understanding consumers' technology acceptance behavior in the setting of 5G. The study aimed to see how consumers' awareness and knowledge influenced their purchasing intentions for 5G technology. To accomplish so, a structured questionnaire was developed and tested in three provinces of Zambia, using previous literature as a guide. The selected six hypotheses were created and analyzed using SEM to discuss the research framework. The numbers, measurement, and structural models fit well, and dependability was also high for various structures.

7.5 Theoretical Implications

According to the findings, customers' WoM and desire to pay for 5G are influenced by their approach toward 5G. As a result, consumers' perceptions of 5G's value and utility improve due to its speed and ubiquity, making them more likely to use it. Privacy issues hamper users' readiness to embrace 5G since they value the security of their personal information. This study discovered the optimistic and positive relationship when combined with reaction time, speed, privacy, ubiquity, and subjective norms to the behavior intention of 5G adoption. More people are ready to use 5G due to their willingness to pay, enhancing their work efficiencies. The chances of promoting 5G over the Internet could be better in Zambia. Furthermore, when customers see 5G as having privacy risks, they negatively view it. The study suggests that 5G telecommunications should ensure privacy and security. Operators of 5G networks should also answer consumers' concerns about the product and resolve the issues raised by the respondents. To boost 5G acceptance, telecommunications firms can also promote a 5G-related initiative through appropriate advertisement design; promotion focuses on consumers. Users would therefore focus on the paybacks of 5G rather than the potential of the technology. The public's perception of 5G would improve their willingness to pay for it, and WoM can grow 5G. Higher willingness to pay may increase the number of 5G customers, resulting in higher adoption and higher acceptability.

7.6 Limitations and Future Research

Despite notable achievement to outline the reality by significant multidimensional relationship and promising results, the model of the present work can still be improved by incorporating more enablers for the adoption of 5G in African countries. There are several flaws and restrictions. First, users' behavior intention was examined in one of the African countries, that is, Zambia. Opinions on 5G were only assessed at the time of the study. As a result, the information presented has less value, so a long-term study with a higher sample size can reduce this constraint. Second, this work solely examined the effects of privacy, speed, ubiquity, and subjective norms on 5G attitudes. Though, the way people think about 5G could be different in other regions of Zambia. While using 5G, several issues include cost-effectiveness, advertisement, policy, and promotion.

A thorough understanding of customer emotions and behavioral characteristics should be included in future research. The past research has stated that organizations with well-structured privacy rules can increase customers' trust (Tang et al., 2021); thus, more research is needed. A small sample of 480 people is improbable to represent the wider public. Nonetheless, the study's findings could be a baseline for future national and international research. Lastly, since the TAM has been widely employed, future studies should consider consumer intention and behavior concerning 5G using alternative theoretical models. Future studies may investigate using the unified theory of acceptance and use of technology (UTAUT) to 5G consumer behavior since it better explains the idea of planned behavior.

7.7 Conclusion

Globalization and technological progression have proposed rapid changes in the telecom sector. Increasing awareness and interest in ICT has augmented demand for better and speedy 5G services. 5G is a technology that provides citizens with faster and more innovative cellular services. While several technological aspects of 5G have been widely studied by experts worldwide, the behavior intention of 5G in terms of business models is lacking in the African continent. Using the TAM, researchers in Zambia examined how privacy risk, subjective standards, speed, and ubiquity affect views toward 5G. 480 respondents evaluated the contributing relationships between the indicated variables and the mediating role of user views regarding 5G. 5G attitudes mediated privacy, speed, ubiquity, subjective criteria, and WoM. The findings could help policymakers, 5G service providers, and suppliers better understand 5G in developing countries like Zambia. The study's value to managers and ICT vendors because 5G initiators often confront difficulties in formulating an appropriate policy for effective planning and development due to the lack of a comprehensive framework of enabler effects of 5G. Thus, academics and practitioners should evaluate these characteristics to understand 5G acceptability and usage better. Manufacturers can more readily complete considerations connected to these issues, including supplying solid infrastructure for a superior 5G service program. This approach can promote technology appreciation and quality of life in civilian society.

References

Ahokangas, P., Matinmikko-Blue, M., Yrjölä, S., & Hämmäinen, H. (2021). Platform configurations for local and private 5G networks in complex industrial multi-stakeholder ecosystems. *Telecommunications Policy*, *45*(5), 102127. https://doi.org/10.1016/j.telpol.2021.102128

Ajzen, I., & Driver, B. L. (1992). Application of the Theory of Planned Behavior to Leisure Choice. *Journal of Leisure Research*, *24*(3), 207–224. https://doi.org/10.1080/00222216.1992.11969889

Chan, W. M., & Lee, J. W. C. (2021). 5G Connected Autonomous Vehicle Acceptance: the Mediating Effect of Trust in the Technology Acceptance Model. *Asian Journal of Business Research*, *11*(1), 1–15. https://doi.org/10.14707/ajbr.210098

Chang, A. (2012). UTAUT and UTAUT 2: A Review and Agenda for Future Research. *The Winners*, *13*(2), 106–114. https://doi.org/10.21512/tw.v13i2.656

Cheng, L., Huang, H., & Yang, S. (2021). Technology in Society Attitude toward 5G : The moderating effect of regulatory focus. *Technology in Society*, *67*, 101795. https://doi.org/10.1016/j.techsoc.2021.101795

Cheng, L. K., Huang, H. L., & Yang, S. Y. (2021). Attitude toward 5G: The moderating effect of regulatory focus. *Technology in Society*, *67*(10), 101795. https://doi.org/10.1016/j.techsoc.2021.101795

Dadhich, M., Hiran, K. K., & Rao, S. S. (2021). *Teaching – Learning Perception Toward Blended E-learning Portals During Pandemic Lockdown*. Springer Singapore. https://doi.org/10.1007/978-981-16-1696-9

Dadhich, M., Purohit, H., & Bhasker, A. A. (2021). Determinants of green initiatives and operational performance for manufacturing SMEs. *Materials Today: Proceedings*, *46*(20), 10870–10874. https://doi.org/10.1016/j.matpr.2021.01.889

Dadhich, M., Rao, S. S., Sethy, S., & Sharma, R. (2021). Determining the Factors Influencing Cloud Computing Implementation in Library Management System (LMS): A High Order PLS-ANN Approach. *Library Philosophy and Practice*, 6281. https://doi.org/https://digital-commons.unl.edu/libphilprac/6281

Davis, F. D.; Bagozzi, R. P.; Warshaw, P. R. (1989). "User acceptance of computer technology: A comparison of two theoretical models. *Management Science*, *35*(8), 982–1003. https://doi.org/10.1287/mnsc.35.8.982, S2CID 14580473

Ding, L., Tian, Y., Liu, T., Wei, Z., & Zhang, X. (2021). Understanding commercial 5G and its implications to (Multipath) TCP. *Computer Networks*, *198*, 108401. https://doi.org/10.1016/j.comnet.2021.108401

Flaherty, E., Sturm, T., & Farries, E. (2022). The conspiracy of Covid-19 and 5G: Spatial analysis fallacies in the age of data democratization. *Social Science & Medicine*, *293*, 114546. https://doi.org/10.1016/j.socscimed.2021.114546

Hair, J. F., Ringle, C. M., Sarstedt, M. (2011). PLE-SEM: Indeed a silverbullet. *Journal of Marketing Theory and Practices*, *19*(2), 139–152.

Hair, J. F., Risher, J. J., & Ringle, C. M. (2018). When to use and how to report the results of PLS-SEM. *European Business Review*, *31*(1), 2–24. https://doi.org/10.1108/EBR-11-2018-0203

Hair Jr, Joseph F., G. Tomas M. Hult, Christian Ringle, and M. S. (2016). *A primer on partial least squares structural equation modeling (PLS-SEM)*

Hair, M, C., DJ, O., & RP, B. (2010). *Essentials of marketing research.*

Hoeschele, T., Dietzel, C., Kopp, D., Fitzek, F. H. P., & Reisslein, M. (2021). Importance of Internet Exchange Point (IXP) infrastructure for 5G: Estimating the impact of 5G use cases. *Telecommunications Policy*, *45*(3), 102091. https://doi.org/10.1016/j.telpol.2020.102091

Hui, H., Ding, Y., Shi, Q., Li, F., Song, Y., & Yan, J. (2020). 5G network-based Internet of Things for demand response in smart grid: A survey on application potential. *Applied Energy*, *257*, 113972. https://doi.org/10.1016/j.apenergy.2019.113972

Jeon, C., Han, S. H., Kim, H. J., & Kim, S. (2020). The effect of government 5G policies on telecommunication operators' firm value: Evidence from China. *Telecommunications Policy*, *1*(May), 102040. https://doi.org/10.1016/j.telpol.2020.102040

Ji, Y., Bai, Y., Liu, X., & Jia, K. (2020). Progress of liquid crystal polyester (LCP) for 5G application. *Advanced Industrial and Engineering Polymer Research*, *3*(4), 160–174. https://doi.org/10.1016/j.aiepr.2020.10.005

Jorquera Valero, J. M., Sánchez Sánchez, P. M., Gil Pérez, M., Huertas Celdrán, A., & Martínez Pérez, G. (2022). Toward pre-standardization of reputation-based trust models beyond 5G. *Computer Standards and Interfaces*, *81*, 103596. https://doi.org/10.1016/j.csi.2021.103596

Kpodar, K., & Fund, I. M. (2014). *ICT, Financial Inclusion, and Growth : Evidence from African Countries Mihasonirina Andrianaivo and Kangni Kpodar* (Issue September). https://doi.org/10.5089/9781455227068.001

Kumar, manish dadhich & naresh. (2015). An Analysis of Factors Affecting to Entrepreneur Development in Rajasthan. *International Journal of Management, IT and Engineering*, *5*(12), 41–48.

Kumar, N., & Dadhich, M. (2014). Risk Management for Investors in Stock Market. *EXCEL International Journal of Multidisciplinary Management Studies*, *4*(3), 103–108.

Kumar Naresh, Dadhich Manish, R. S. S. (2014). Determinant of Customers' Perception towards RTGS and NEFT Services. *Asian Journal of Research in Banking and Finance*, *4*(9), 253–260. https://doi.org/10.5958/2249-7323.2014.00960.2

Manvinder Singh Pahwa, Manish Dadhich, Jaskaran Singh Saini, D. kumar S. (2021). Use of Artificial Intelligence (AI) in the Optimization of Production of Biodiesel Energy. In *Artificial Intelligence for Renewable Energy Systems* (p. 230). Scrivener Publishing, Wiley. https://doi.org/doi.org/10.1002/9781119761686.ch11

Naresh Kumar, M. D. (2016). An analytical study of life insurance facilities provided by life insurance companies. *SAARJ Journal on Banking & Insurance Research*, 5(1), 82–92.

Nicolas, C., Kim, J., & Chi, S. (2020). Quantifying the dynamic effects of smart city development enablers using structural equation modeling. *Sustainable Cities and Society*, 53, 101916. https://doi.org/10.1016/j.scs.2019.101916

Oughton, E. J., Lehr, W., Katsaros, K., Selinis, I., Bubley, D., & Kusuma, J. (2021). Revisiting Wireless Internet Connectivity: 5G vs Wi-Fi 6. *Telecommunications Policy*, 45(5), 102127. https://doi.org/10.1016/j.telpol.2021.102127

Parcu, P. L., Innocenti, N., & Carrozza, C. (2021). Ubiquitous technologies and 5G development. Who is leading the race? *Telecommunications Policy*, 1(10), 1–15. https://doi.org/10.1016/j.telpol.2021.102277

Rajapakshe, W. (2021). The Role of Micro Finance Institutions on the Development of Micro Enterprises (MEs) in Sri Lanka. *South Asian Journal of Social Studies and Economics*, 9(1), 1–17. https://doi.org/10.9734/sajsse/2021/v9i130227

Rakesh Kumar Birda & Manish Dadhich. (2019). Study of ICT and E-Governance Facilities in Tribal District of Rajasthan. *ZENITH International Journal of Multidisciplinary Research*, 9(7), 39–49.

Rana, A., Taneja, A., & Saluja, N. (2021). Accelerating IoT applications new wave with 5G: A review. *Materials Today: Proceedings*, 10(1), 1–12. https://doi.org/10.1016/j.matpr.2021.03.292

Rogalski, M. (2021). Security assessment of suppliers of telecommunications infrastructure for the provision of services in 5G technology. *Computer Law and Security Review*, 41, 105556. https://doi.org/10.1016/j.clsr.2021.105556

Rossi, M. A. (2021). The advent of 5G and the non-discrimination principle. *Telecommunications Policy*, 3, 102279. https://doi.org/10.1016/j.telpol.2021.102279

Shah, S. K., Zhongjun, T., Sattar, A., & XinHao, Z. (2021). Consumer's intention to purchase 5G: Do environmental awareness, environmental knowledge and health consciousness attitude matter? *Technology in Society*, 65, 101563. https://doi.org/10.1016/j.techsoc.2021.101563

Shalendra Singh Rao, M. D. (n.d.). Impact of Foreign Direct Investment in Indian Capital Market. *International Journal of Research in Economics and Social Sciences (IJRESS)*, 7(6), 172–178.

Sharma, N., & Dadhich, M. (2014). Predictive Business Analytics: The Way Ahead. *Journal of Commerce and Management Thought*, 5(4), 652. https://doi.org/10.5958/0976-478x.2014.00012.3

Singh, G. K. (2021). Impact Execution of Total Quality Management (TQM) on Operational Performance of Indian Cement Manufacturing Industry: A Comprehensive SEM Approach. *Design Engineering*, *8*, 13538–13562. https://doi.org/http://thedesignengineering.com/index.php/DE/article/view/6476

Siwale, J. (2021). Microfinance regulation and social sustainability of microfinance institutions : The case of Nigeria and Zambia. *Annals of Public and Cooperative Economics*, *1*, 1–30. https://doi.org/10.1111/apce.12165

Sreelakshmi Krishnamoorthy, Amit Dua, S. G. (2021). Role of emerging technologies in future IoT-driven Healthcare 4.0 technologies a survey, current challenges and future directions. *Journal of Ambient Intelligence and Humanized Computing*, *1*, 1–15. https://doi.org/10.1007/s12652-021-03302-w

Tang, Y., Dananjayan, S., Hou, C., Guo, Q., Luo, S., & He, Y. (2021). A survey on the 5G network and its impact on agriculture: Challenges and opportunities. *Computers and Electronics in Agriculture*, *180*, 105895. https://doi.org/10.1016/j.compag.2020.105895

8

Study of Cybersecurity in 5G using a Machine Learning Algorithm for Protecting Financial Transactions – A Case of the Developing World

Shubham Goswami[1], Bhawna Hinger[2], Vineet Chouhan[3], and Tarannum Hussain[4]

[1]Sir Padampat Singhania University, India
[2]Govt. Meera (Girls) College, India
[3]Sir Padampat Singhania University, India
[4]Mohanlal Sukhadia University, India
Email: shubham.goswami@spsu.ac.in; vcpc2008@gmail.com

Abstract

5G will provide quick payment alternatives, making digital payments even more attractive to consumers and businesses. With appropriate deployment, 5G will fasten contemporary financial services using next-generation cloud-computing operations and AI. But establishing 5G mobile networks in developing countries may be challenging. The many security concerns of 5G networks are continuing issues, and the developing world is not technologically capable of handling them. With the rapid growth of e-commerce in developing nations, transaction fraud has increased dramatically, as more people use online services to conduct their daily transactions. Illegal online transactions have become more common worldwide, resulting in significant financial losses for most enterprises and individuals. Fraud trends are started evolving, requiring a shift from reactive to proactive fraud detection. Financial fraud is quite likely in today's fast-paced world when billions of financial transactions are made daily. Fraud trends are shifting, necessitating a shift from a reactive to a proactive strategy to fraud detection. Because of the dynamic nature of fraudulent conduct, which is scattered over multiple customer profiles and dispersed in massive and active datasets, online bank fraud is difficult to evaluate and identify. The current chapter employs supervised learning methods such as logistic regression (LR), decision tree, and artificial neural network (ANN) to construct an effective model for more accurate classification and prediction with labeled data. It compares the accuracy and efficiency of the various approaches.

8.1 Introduction

With the frequent expansion of e-commerce, transaction fraud has expanded exponentially, where people are using more online services to perform their daily transactions. The reliance on e-commerce and online payments has increased steadily over the past several decades. As the field of information technology improves day by day, unlawful efforts in online transactions have grown globally, resulting in significant financial losses for most businesses and individuals. First, fraud trends are evolving, necessitating a shift from a reactive to a proactive strategy to fraud detection. Financial fraud is highly likely in today's fast-paced world, where billions of financial transactions occur daily. Financial Fraud Action UK (FFA UK) estimated £755 million in damages in 2015, a 26% increase over 2014. Fraud losses on UK-issued cards reached £567.5 million in 2015, an 18% rise from £479 million the previous year: the fourth consecutive year of growth (Flatley, 2016).

Approximately $20 billion was stolen in the United States in the same period due to credit card theft, with 12.7 million victims. According to this data, credit card fraud is a global issue affecting financial institutions and consumers. Fraud may be committed in a variety of ways (Irvin-Erickson & Ricks, 2019). A stolen identity is one method in which a fraudster creates a fake id and injects it into the bureau to issue a new credit card (Aïmeur&Schönfeld, 2011). Due to the dynamic nature of fraudulent activity distributed over various customer profiles and dispersed in massive and active datasets, online bank fraud is difficult to evaluate and identify. Algorithms and analytical technologies have been used to create complex decision-making systems. These may learn from past events and develop patterns that can identify possible fraudulent transactions ahead of time (Loginovsky etal., 2020).

As the fraudulent activity is dynamic, distributed across various client profiles, and scattered in massive and active datasets, online bank fraud continues to evolve and is challenging to evaluate and identify. Algorithms and analytical technologies have been used to create complex decision-making systems (Goswami and Chouhan, 2021; Khan etal., 2014). These may learn from past events and develop patterns that can identify possible fraudulent transactions ahead of time.

Online banking fraud has become a significant problem for all banks (Chouhan et.al., 2020a). The growth of sophisticated online financial fraud schemes, including phishing, malware, and ghost websites, is becoming more complex and results in huge losses. Online banking fraud must be detected immediately since it is complicated to recover losses if the copy is not found within the detection period (Chouhan et al., 2021c). Most consumers seldom check their online banking history regularly, and as a result, they are unable to detect and report fraudulent transactions soon after they occur. As a result, the chances of recouping a loss are very slim. Detection systems used for online banking should have extreme accuracy, detection rate, and small untrue optimistic rate to provide a fair number of warnings in a complicated online banking environment (Wei et al., 2013). These characteristics significantly complicate current fraud detection techniques for credit card transactions, e-commerce, insurance, retail, telecommunications, computer intrusion, and other sectors. Many existing methods perform poorly in efficiency and accuracy (Chouhan et al., 2021b).

These con artists create a fake credit history and defraud numerous banks. It is impossible to manually identify fraudulent transactions among millions of online transactions occurring in a fraction of a second, necessitating automated fraud detection system. Machine learning offers automatic and faster detecting of fraud in millions of transactions without requiring human involvement (Levi, 2014). Credit card fraud detection is a binary classification issue, with the transaction being either fraudulent or legal as a result. When analyzing transactions, there are three key points to keep in mind. Due to the difficulty of detecting fraudulent transactions in a highly unbalanced dataset, an innovative mining solution must balance the dataset without sacrificing any essential information (Zhang etal., 2019).

The second aspect is to develop a machine learning method that can accurately learn from such an unbalanced dataset. False positives result in a smaller financial loss for the business than false negatives (fraudster recognized as genuine), resulting in significant economic losses in a short period (Nesvijevskaia etal., 2021). Consequently, the third important aspect is to evaluate the outcomes of trained models using the best model performance measures. Bagging, boosting, and stacking are three popular ensemble machine learning techniques used in fraud detection, in addition to sampling approaches. These ensemble machine learning techniques combine several learning algorithms to provide more excellent prediction performance than any individual learning algorithm could provide (Chouhan et al., 2021d; Dietterich, 2020).

Financial crime management has grown more concerned about online financial fraud. It is becoming more complex and there are substantial losses due to sophisticated and creative online banking fraud, including phishing schemes, malware infection, and ghost websites. Because it is challenging to recover damages if fraud is not found within the detection period, online banking fraud must be caught promptly (Surana et al., 2021). In addition, credit card fraud is a global problem that affects both financial institutions and consumers in most developing nations. To generate a few alarms in a complicated online banking environment, online fraud detection systems should have high accuracy, a high detection rate, and a low false-positive rate. These characteristics significantly challenge existing fraud detection methods in credit card transactions, e-commerce, insurance, retail, telecommunications, computer intrusion, and other sectors. These current methods are inefficient for timely detection of online financial fraud. Machine learning provides an automated and quicker process to identify fraud in millions of transactions without human involvement. In general, fraud detection is a binary classification issue, with the transaction being classified as either fraudulent or legitimate.

8.2　Relevance of Financial Security in 5G

5G will provide quicker and more manageable payment alternatives, making mobile and digital expenditures even more attractive to consumers and businesses alike, increasing use. This is crucial for economic development. Consumers increased their use of digital payments due to the COVID-19 pandemic since it provided a contactless transaction method and unmatched convenience during the lockdown time. As a consequence, businesses – mainly e-commerce – received a once-in-a-lifetime boost. With the assistance of 5G, economic institutes may act significantly in developing this industry in 2021 (Balaraman, 2021).

Another significant benefit of 5G is allowing banks to enhance proactive fraud protection and make educated real-time choices. Incorporating client geolocation and merchant ID will minimize fraud detection mistakes and false positives, safeguarding customers and the bank's bottom line (Jiang, 2021; Raj & Raman, 2015).

Since 5G devices still link to older networks, security flaws persist. Due to a weakness in the GSMA standard, individuals may still be monitored when connecting to 5G using information that stays unencrypted while it is sent. 5G provides a few paramount privacy and security advantages: it encrypts identifiers and includes security and privacy capabilities like spoofing and anti-tracking (Khan & Martin, 2020; Catania, 2019). These features may offer substantial benefits, such as safeguarding users against online network threats and manipulation. 5G, on the other hand, comes with its own set of security issues. It is anticipated to accelerate the development of mobile Internet use for both individuals and businesses, resulting in an explosion of IoT-connected devices. With appropriate deployment, 5G will hasten contemporary financial services, transforming ordinary activities like collecting credit information into next-generation edge- and cloud-computing operations, AI, and more. In terms of the latter, numerous financial innovations are already reducing the importance of traditional banking. During COVID-19, the demand has only grown. A growing number of financial institutions are focusing on creating a frictionless, secure, quick, and simple client experience. These functionalities will become more common thanks to 5G.

The explosion of connected devices and sensors will open up many new possibilities to profit from data and data intelligence. However, the increasingly deployed devices also provide a significant outbreak surface for impostors to takeover IoT strategies and launch DDoS assaults. Hackers are breaking into IoT devices at an alarming rate, making fraudulent transactions and launching DDoS assaults. While hackers continue to rob Content Security Policy (CSP) and their customers of billions of dollars each year via text messaging, voice calling systems, and IP exchange, new types of SIM card theft are increasing. One of the telecom industry's most outstanding fraud issues is "SIM boxing". Fraudsters utilize SIM used for making international calls seem to be local ones. Fraudsters pose as legal intermediaries, but they route calls across international boundaries to "SIM boxes", which look like a cluster of local mobile phones.

Telecom operator's loss in millions of dollars of revenue each year is a consequence of missed international calls. Governments lose tax money as a result, while end-users get poor service. A whole ecosystem has grown around it to enable this, with more complex software being used to evade detection. In reaction to countermeasures, fraudsters also alter their conduct quickly. In terms of fraud detection and prevention, in-event detection and prevention are much more effective and desired than post-event reconciliation. The subscriber's (or CSP's) credit card has been already charged. The revenue has indeed been lost because of the fraud.

In-event fraud detection and prevention, on the other hand, is far easier said than done at the size and speed of 5G. Simply stated, conventional fraud protection methods used by CSPs will no longer be effective in the 5G future. Companies will need a data management platform that can quickly ingest data from many sources and apply hundreds of "machine-learned" rules to fight fraud (i.e., under 10 milliseconds).

Because of 5G's enhanced capacity, millions of sensors and cameras will be able to connect, allowing for a vast consumer data-gathering network. Because of more comprehensive consumer monitoring, service optimization and personalization will be feasible. This will enhance each bank's duty to safeguard client data, which will almost certainly extend beyond just ticking the boxes on the General Data Protection Regulation (GDPR) and the California Consumer Privacy Act (CCPA).

8.3 Review of Literature

Online-banking fraud has developed as a significant problem for all banks. Due to the growth of erudite online financial fraud schemes, including phishing, malware, and ghost websites, is becoming more complex and results in huge losses. Online banking fraud must be detected immediately since it is complicated to recover losses if the copy is not found within the exposure time (Chouhan et al., 2021e). Most consumers seldom check their online banking history regularly, and as a result, they are unable to detect and report fraudulent transactions immediately after they occur. As a result, the chances of recouping a loss are very slim. In this respect, Internet banking detection methods should be very accurate, have a high detection rate, and have a low false-positive rate to produce a restricted, reasonable number of notifications in online banking. These characteristics are seriously challenged by techniques for detecting card fraud, e-commerce, healthcare, retail, communications, computer intrusion, and other sectors. When used directly to online banking fraud detection, several existing methods operate inefficiently and inaccurately. Credit card fraud detection, for example, often depends on recognizing the behavior patterns of a particular customer or group. Simultaneously, fraud-related online banking transactions are fluid and seem to be very comparable to actual customer behavior. In a dynamic computing environment, specific intrusion detection techniques work well, but they need a significant quantity of training data and full attack logs as proof. There is, however, no clear proof that an online banking transaction is fraudulent.

Elena Adriana Minastireanu (2019), in their study on "An Analysis of the Most Used Machine Learning Algorithms for Online Fraud Detection", attempted to uncover the most common problems that credit card fraud detection methods face. The supervised learning techniques: Automated Valuation Models (AVM), ANN, and decision tree produced the most outstanding results in terms of accuracy and coverage, according to the categorization of the algorithms. It would be beneficial to look at how algorithms might be improved to make them more applicable to various online fraudulent transactions with high accuracy, coverage, and cheap costs.

Davis Robertson (2016), in his paper on "Top Card Issuers in Asia-Pacific Card Fraud Losses Reach", studied the imbalanced and skewed distribution of data. The researcher found that data mining methods take time to perform when dealing with large amounts of data. However, the researcher shows the behavioral aspect of big data. Thus this research proposal will provide a technological solution with an experimental study.

Thennakoon et al. (2019), in their paper on "Real-time Credit Card Fraud Detection Using Machine Learning", through a comparison of machine learning methods, the best algorithms for the four fraud patterns were found. They did, however, suggest a new credit card fraud detection system. As demonstrated in the approach, optimal algorithms that handle four significant frauds were chosen via reading, testing, and parameter tweaking.

ThulaSyammalRamiah Pillai (2018), in their paper on "Credit Card Fraud Detection Using Deep Learning Technique", used deep learning methods to improve accuracy. It is the most effective method, and it is utilized in various applications such as voice recognition. The MLP machine learning algorithm was used. MLP necessitates fine-tuning all parameters, including the number of hidden layers, neurons, and different kinds of activation functions. According to the findings, MLP with a logistic activation function produces the best outcomes.

Kuldeep Randhawa (2017), in their paper on "Credit card fraud detection using AdaBoost and majority voting", 12 conventional models and hybrid approaches using AdaBoost and public voting was used to improve fraud detection accuracy. It used both benchmarking and real-world data. The methods' benefits and disadvantages were reviewed. The MCC was employed as a performance indicator. This was done to see how well the algorithms handled noise. They showed that the extra noise did not affect majority voting. A real-world credit card data collection from a bank was also used for the study.

Brause, Rüdiger, T. Langsdorf, and Michael Hepp (1999), in their paper on "Neural data mining for credit card fraud detection", revealed that credit card fraud prevention is a significant use for prediction methods. One significant impediment to utilizing neural network training methods is the high diagnostic quality required. Since just one financial transaction out of a thousand is incorrect, a forecast success rate of less than 99.9% is unacceptable. Due to the credit card transaction requirements, entirely new ideas had to be created and validated using actual credit card data. This article demonstrates how sophisticated data mining methods and a neural network algorithm may be effectively coupled to provide high fraud coverage while minimizing false alarms.

Ghosh, Sushmito, and Douglas L. Reilly (1994), in their paper on "Credit card fraud detection with a neural-network", revealed that they built a neural network-based fraud detection system using credit card transaction data from a credit card company. The neural network was trained using lost, stolen, and nonreceived issue (NRI) cards. Overrule-based fraud detection methods discovered significantly more fraudulent accounts while producing far fewer false positives (by a factor of 20). Mellon Bank has placed the system on an IBM 3090 and is now using it to identify fraud on its credit card portfolio. We evaluate the network's accuracy and speed in identifying fraud utilizing this dataset.

Chan, Philip K., Wei Fan, Andreas L. Prodromidis, and Salvatore J. Stolfo (1999), in their paper on "Distributed data-mining in credit card fraud-detection", revealed that credit card transactions are increasing in volume, accounting for a more significant proportion. Enhanced fraud detection is critical for the payment system to thrive in America. Banks have long utilized early fraud detection systems. Large-scale data mining has the potential to improve corporate practices. Scalable methods for analyzing large quantities of transaction data and effectively computing fraud detectors quickly, particularly for e-commerce, are critical issues. In addition to scalability and efficiency, the fraud-detection job has technical problems. The knowledge discovery and data mining groups have paid little attention to skewed training data distributions and nonuniform cost per error. In this article, we examine and assess various methods that simultaneously solve these three major problems. Our suggested techniques for integrating several trained fraud detectors under a "cost model" are generic and demonstrably effective; our empirical findings show that distributed data mining of fraud models may substantially decrease loss due to fraud.

Srivastava, Abhinav, Amlan Kundu, Shamik Sural, and Arun Majumdar (2008) in their paper on "Credit card fraud-detection using hidden Markov model" explained that the usage of credit cards had risen significantly as a result of fast advancements in electronic commerce technologies. As credit cards become the most common payment method for online and offline purchases, credit card fraud is on the rise. In this article, we utilize a hidden Markov model (HMM) to describe the sequence of credit card transaction processing processes and demonstrate how it may be used to identify fraud. An HMM is first trained on a cardholder's typical behavior. It is fraudulent if the trained HMM does not accept an incoming credit card transaction with a high enough probability. At the same time, we strive to avoid rejecting legitimate transactions. We provide comprehensive experimental data to demonstrate our method's efficacy and compare it to other processes in the literature.

Quah, Jon TS, and M. Sriganesh (2008) in their paper on "Real-time credit card fraud detection using computational intelligence", explained that online banking and e-commerce had seen significant development in recent years, and they continue to show great potential for the future. It has finished it simpler for criminals to engage in new and complex methods of credit-card fraud over the Internet. This article focuses on real-time fraud-detection and offers a novel approach to decipher expenditure patterns to identify probable fraud instances. For fraud detection, it employs a self-organization map to interpret, filter, and analyze consumer activity.

Dal Pozzolo et al., (2014) in their work "Learned lessons in credit card fraud detection from a practitioner perspective", expressed that fraudulent credit card transactions result in billions of dollars in losses every year. The development of practical fraud detection algorithms is critical for minimizing these losses, and more algorithms are turning to sophisticated machine learning methods to help fraud investigators. Designing fraud detection algorithms is especially difficult because of the nonstationary distribution, extremely unbalanced class distributions, and continuous streams of transactions. At the same time, public data is limited due to confidentiality concerns, leaving many questions unresolved regarding the appropriate approach for dealing with them. This article concentrates on three critical issues: imbalanced, non-stationarity, and evaluation, and offer some solutions from a practitioner's viewpoint. A genuine credit card dataset supplied by our industrial partner enabled the research.

Shen, Aihua, Rencheng Tong, and Yaochen Deng (2007), in their research on "Application of classification models on credit card fraud detection", revealed that credit card fraud has grown more common in recent years, owing to the significant rise in credit card transactions. This research looks at the effectiveness of using classification models to solve credit card fraud detection issues. The usefulness of three distinct classification techniques, namely decision trees, neural networks, and LR, in fraud detection is investigated. Their article lays forth a helpful approach for determining the optimal model for detecting credit card fraud.

Aleskerov, Emin, Bernd Freisleben, and Bharat Rao (1997) in their paper on "Cardwatch: A neural network-based database mining system for credit card fraud detection", CARDWATCH, a database mining method for credit card fraud detection, is demonstrated and described. The learning module built by the system is on a neural network, includes a graphical user interface, and connects to several commercial databases. The tests using synthetically produced credit card data and an auto-associative neural network model indicate high fraud detection rates.

Raj, S. Benson Edwin, and A. Annie Portia (2011), in their paper on "Analysis on credit card fraud detection methods", expressed that due to the fast development of e-Commerce, credit card fraud has grown significantly. Credit card fraud is on the rise as credit cards become the primary payment method for online and offline purchases. In real life, fraudulent transactions coexist alongside legitimate transactions, and basic pattern matching techniques are often insufficient to identify scams correctly. To reduce their losses, all credit card issuing institutions must have effective fraud detection systems. Many contemporary methods for identifying different credit card fraudulent transactions have developed, including AI, data mining, fuzzy logic, ML, sequence alignment, genetic programming, and others. Thorough knowledge of these methods will undoubtedly result in a very effective credit card fraud detection system.

8.4 Research Problem

Because the developed ML models are only averagely accurate, we must focus on improving prediction levels. When it comes to transaction analysis, there are three significant issues. First, finding fraudulent transactions in a highly imbalanced dataset is difficult; intelligent mining solutions are needed to balance the dataset without losing any valuable information. Traditional data mining techniques take longer to find accuracy.

False positives result in a minor financial loss for the company than false negatives (fraudster identified as genuine), resulting in significant economic losses in a short period. The secondary problem is to create a machine learning method that can accurately learn from such an unbalanced dataset. As a result, the third important aspect is to examine the output of trained models using the best model performance measures. Bagging, boosting, and stacking are three typical ensemble machine learning techniques used in fraud detection, in addition to sampling methods. These ensemble machine learning techniques integrate several learning algorithms to offer better forecasting performance than any individual learning algorithms could provide.

8.5 The Objective of the Chapter

On the academic front, issues like fraud detection and prevention have gotten much attention from payment card issuers in recent years. The present research can be attributed to the significant annual financial losses incurred by fraudulent use of their financial products. This chapter aims to create a model that can forecast the future. Financial transactions that could be fraudulent can be detected with high accuracy. Based on the literature review, the most often used method for identifying fraudulent transactions seems to be supervised learning algorithms. It also claims that the most common dishonest e-commerces are those conducted using a credit card. Another goal of this study is to develop a model to identify fraud using machine learning techniques. The model developed will alert the system and witness the most prominent fraud demographics like location. The study further analyzes and predicts fraud in transactions based on a machine learning algorithm. Since the input is a large file of unstructured transactions, we perform preprocessing, extract features on which predictions are made, and then classify as fraud or legitimate transactions.

Payment card issuers have paid a lot of attention to themes like fraud detection and prevention on the research front in recent years. The current chapter is motivated by the significant annual financial losses incurred due to fraudulent usage of their financial products. The aim is to build a model that can anticipate fraudulent financial transactions with high accuracy. The objective is to use a machine learning system to analyze and predict fraud in commerce. The current chapter employs supervised learning methods such as LR, random forest, and support vector machine (SVM) to construct an efficient model for more accurate classification and prediction with labeled data. It compares the accuracy and efficiency of the various approaches.

8.6 Methodology

Machine learning is a branch of artificial intelligence that combines artificial intelligence, mathematics, psychology, neuroscience, and information technology. As a result, ML algorithms are divided into two types: supervised learning algorithms (artificial immune system, ANN, Bayesian network (BN), SVM, decision tree, LR, Naive Bayes, random forest, fuzzy logic-based system, and K-nearest neighbor) and unsupervised learning algorithms (artificial immune system, ANN, BN, SVM, decision tree, LR, Naive Bayes (hidden Markov Model, self-organizing map, genetic algorithm, K-means, DBSCAN, expert system, gradient descendent, and scatter search).

Supervised techniques offer excellent accuracy and coverage but come at a high cost. Due to lack of consistency in the dataset, they were highly unbalanced, with many negative occurrences, it is critical to compare the results. The majority of transaction data attributes have categorical values.

ML algorithms virtually never support absolute values. As a result, detecting techniques and feature selection are brutal to recognize. Another significant issue in financial fraud detection is feature selection. It tries to pick out the traits that best characterize the qualities and aspects of fraud.

The present research use supervised learning algorithms for more accurate classification and prediction with labeled data. The review also suggests that the fraud detection system is based on SVM, BN, fuzzy logic, and DBSCAN with high accuracy when processing large datasets.

- Understanding the data for preprocessing and cleaning of datasets is a critical task. The data will be collected, including financial transactions. To prevent the risk of learning wrong data patterns, a common transform technique is principal component analysis. From the standpoint of numerical analysis, this approach addresses the feature selection problem. PCA will be applied to the dataset for choosing an appropriate number of principal components.

- In the data cleansing process, it's critical to fill in missing values. There are numerous solutions, such as disregarding the entire tuple. An appropriate technique like null value replacement, average value replacement, interpolation\extrapolation will be applied after analysis of descriptive statistics of the data.

- The number of instances of fraudulent transactions was considerably lower than the number of examples of legitimate commerce. To address this, the researcher will use undersampling and oversampling to reduce majority occurrences while increasing minority occurrences. SMOTE (Synthetic Minority Oversampling Methods) and closest neighbor undersampling techniques will be employed based on the data properties.

- The data needed to create the final model is typically derived from several different sources. It's crucial to divide the data into training and testing sets. The training dataset is a collection of instances used to fit the model's parameters. In contrast, the test dataset is a collection of cases used to objectively assess the final model fit on the training dataset. The present research uses the famous rule of splitting 80–20% training and testing sets, respectively.

- The ability to precisely classify observations is precious. The present study will apply selected supervised learning classifiers to our re-sampled data. The researcher will apply ML algorithms like LR, random forest, and

SVM for developing an efficient model. LR is used for predictive analysis and describes the connection between a nominal, ordinal, interval, or ratio-dependent binary variable and one or more independent variables. Random forest (RF) is an ensemble learning technique for classification, regression, and other problems. In the present study, RF will be used for classification and regression. RF gives the best accuracy compared to other techniques like decision tree classification tools. It avoids over-fitting by integrating the results of several decision trees.

- The SVM is a machine type that may be used for both regression and classification. However, it is frequently employed in categorization goals. For imbalanced data, the SVM serves as a preprocessor. SVM produces additional data for the minority class. Multiple classification methods are trained using the changed training data. The SVM algorithm aims to identify a hyperplane in N-dimensional space (N – the number of features) that distinguishes between data points.

- Statistical software and data analysis are often developed using the R language, which statisticians and data miners extensively use. Programming, converting, finding, modeling, and communicating the results are all stages in the R data analysis process. The use of machine learning algorithms and data visualization are the most common benefits of R. R-studio will be used in the study to code in R and evaluate the suggested model. Tableau will be utilized for data visualization, such as generating geographic graphs and heat maps to show where credit card fraud occurs.

- With their comparative analysis, these algorithms' performance is validated and compared. The confusion matrix is used to assess the performance of the algorithms. The confusion matrix is the most straightforward technique for evaluating performance since it allows you to see how many data instances are correctly categorized.

8.7 Result and Discussion

For this credit card fraud classification problem, we are using the dataset downloaded from the Kaggle platform. First, data scaling is performed. Scaling is sometimes referred to as feature standardization. The data is organized according to a defined range with the assistance of scaling. As a result, there are no life-threatening values in the dataset that may interfere with the operation of our model. With a split ratio of 0.80, the dataset is divided into a

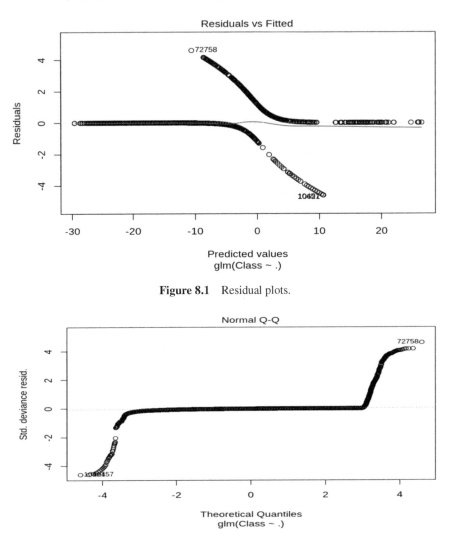

Figure 8.1 Residual plots.

Figure 8.2 Q-Q Plot.

training set and a test set. This implies that 80% of the data will be assigned to the train data, while 20% will be allocated to the test data.

Later the testing of the model begins with LR. It is used for modeling the significance of the likelihood of a class such as positive or negative and, in our case – fraud or not a fraud. The following are the plots for LR including Residuals Plot (Figure 8.1), Q-Q Plot (Figure 8.2) with Pridicted value plot (Figure 8.3).

ROC stands for optimistic receiver characteristics. It is a graphical representation of a binary classifier system's diagnostic performance as its

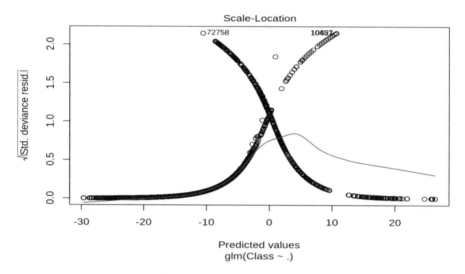

Figure 8.3 Predicted value plot.

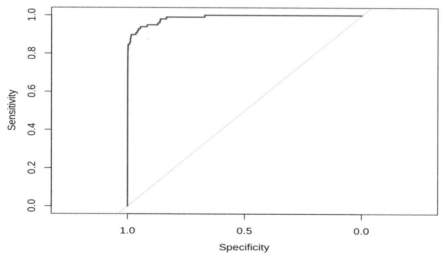

Figure 8.4 ROC plot.

discriminating threshold changes. The ROC curve is displayed to evaluate the model's performance (Figure 8.4).

Next, a **Decision Tree (DT)** process is implemented. It is a plot with the consequences of a pronouncement. These consequences are a result of which it is possible to determine which class the object relates to and results of this classifcation is shown in Figure 8.5.

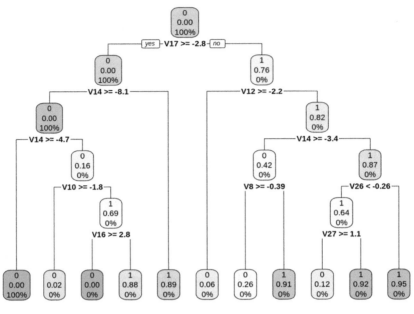

Figure 8.5   Decision tree.

At last, a machine learning algorithm based on the human nervous system ANN models may learn forms from previous data and conduct sorting on the input information. In the case of ANN, there is a value range between 1 and 0. We established a threshold of 0.5; values greater than 0.5 resemble 1, and the remainder is 0. The final ANN for the given study is represented in Figure 8.6.

We used machine learning to create a credit card fraud detection model. We implemented this model using several machine learning methods, displayed the models' relative performance curves, and analyzed and visualized fraudulent transactions from various kinds of data.

8.8 Conclusion

Fraudulent transactions or activities are a significant problem in many sectors, such as banking and insurance, and they are also crucial to the upcoming 5G network. Credit card fraud detection is a severe problem for the banking sector, in particular. These sectors have suffered a great deal due to fraudulent actions aimed at increasing income and losing customers' confidence. As a result, these businesses must identify fraudulent transactions before they become a significant issue. In contrast to other machine learning tasks, the target class

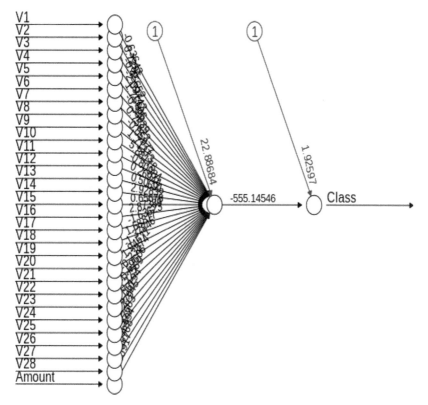

Figure 8.6 ANN model.

distribution in credit card fraud detection is not evenly distributed. It's also known as the imbalanced data problem or the class imbalance problem. Fraud detection with ML becomes possible due to the ability in the 5G network; ML algorithms will acquire past scam tendencies and recognize them for the following transactions. When rapidity matters in which information is processed, machine learning algorithms seem more effective than people.

Furthermore, machine learning algorithms can identify complex fraud characteristics that a person cannot. Rule-based fraud protection systems include writing precise textual rules that "tell" the algorithm which kinds of activities seem regular and should be allowed and which should not because they appear suspect. Writing regulations, on the other hand, takes a long time.

5G is anticipated to enable large-scale microservices such as IoT. The database engine must apply hundreds of rules using machine learning in milliseconds while executing millions of operations every second to prevent fraud. ML presented higher accuracy and size of the dataset used to fit with higher fraudulent operations. For example, they have been qualified, the better they will be at detecting fraud. If rule-based systems never develop, this concept does not apply to them. A team of data scientists should also be mindful of dangers associated with rapid modeling scale; if it fails to identify deception and mismarked it, it may result in subsequent false negatives. Machines will take over primary duties and the tedious labor of manual fraud investigation, allowing experts to focus on higher-level judgments.

Both AI and ML can play a major role in developing, modeling, and automating effective security procedures against a varied and wide variety of threats. If the current era is flooded with card-not-present dealings online, the related sectors will be jeopardized and have many cases of deception. Others are all instances of illegal attacks on vulnerable users' data that result in data breaches. New top-notch techniques based on machine learning algorithms for fraud detection and prevention offer more outstanding value to companies with their real-time work, speed, and efficiency while previous rule-based algorithms fade into the past. In recent years, AI and machine learning developments have had the potential to improve the performance of next-generation 5G networks. AI and machine learning have paved the path for more robust and dynamic solutions in 5G security, privacy, and threat detection.

References

Aïmeur, E., & Schönfeld, D. (2011, July). The ultimate invasion of privacy: Identity theft. In *2011 Ninth Annual International Conference on Privacy, Security and Trust* (pp. 24–31). IEEE.

Aleskerov, E., Freisleben, B., & Rao, B. (1997, March). Cardwatch: A neural network based database mining system for credit card fraud detection. In *Proceedings of the IEEE/IAFE 1997 computational intelligence for financial engineering (CIFEr)* (pp. 220–226). IEEE.

Balaraman, M. (2021). *Customer to Human: The CX Factor in Modern Business*. Penguin Random House India Private Limited.

Brause, R., Langsdorf, T., & Hepp, M. (1999, November). Neural data mining for credit card fraud detection. In *Proceedings 11th International Conference on Tools with Artificial Intelligence* (pp. 103–106). IEEE.

Chan, P. K., Fan, W., Prodromidis, A. L., & Stolfo, S. J. (1999). Distributed data mining in credit card fraud detection. *IEEE Intelligent Systems and Their Applications*, *14*(6), 67–74.

Chouhan, V. (2016). Investigating Factors Affecting Electronic Word-of-Mouth. In *Capturing, Analyzing, and Managing Word-of-Mouth in the Digital Marketplace* (pp. 119–135). IGI Global.

Chouhan, V., Goswami, S., & Sharma, R. B. (2021). Use of proactive spare parts inventory management (PSPIM) techniques for material handling vis-à-vis cement industry. *Materials Today: Proceedings, 45,* 4383–4389.

Chouhan, V., Goswami, S., Dadhich, M., Saraswat, P., & Shakdwipee, P. (2021). Emerging opportunities for the application of blockchain for energy efficiency. *Blockchain 3.0 for Sustainable Development, 10,* 63.

Chouhan, V., Sharma, R. B., Vasita, M. L., & Goswami, S. (2021). Measuring barriers in adoption of blockchain in supply chain management system. *Blockchain 3.0 for Sustainable Development, 10,* 37.

Chouhan, V., Vasita, M. L., & Goswami, S. (2020). The impact and role of social media for consciousness of COVID-19 pandemic. *Journal of Content, Community and Communication, 12,* 250–262.

Dal Pozzolo, A., Caelen, O., Le Borgne, Y. A., Waterschoot, S., & Bontempi, G. (2014). Learned lessons in credit card fraud detection from a practitioner perspective. *Expert systems with applications, 41*(10), 4915–4928.

Demla, N., & Aggarwal, A. (2016). Credit card fraud detection using SVM and Reduction of false alarms. *International Journal of Innovations in Engineering and Technology, 7*(2), 176–182.

Dheepa, V., & Dhanapal, R. (2012). Behavior based credit card fraud detection using support vector machines. *ICTACT Journal on Soft computing, 2*(4), 391–397.

Dietterich, T. G. (2000, June). Ensemble methods in machine learning. In *International workshop on multiple classifier systems* (pp. 1–15). Springer, Berlin, Heidelberg.

Dighe, D., Patil, S., & Kokate, S. (2018, August). Detection of credit card fraud transactions using machine learning algorithms and neural networks: A comparative study. In *2018 Fourth International Conference on Computing Communication Control and Automation (ICCUBEA)* (pp. 1–6). IEEE.

Flatley, J. (2017). Crime in England and Wales: year ending Sept 2016. *London: Office for National Statistics–Crime Survey of England and Wales (CSEW).*

Ghosh, S., & Reilly, D. L. (1994, January). Credit card fraud detection with a neural-network. In *System Sciences, 1994. Proceedings of the Twenty-Seventh Hawaii International Conference on* (Vol. 3, pp. 621–630). IEEE.

Gurram, SK, Naidu, MV (2019). Credit Card Fraud Detection System Based on Machine Learning Techniques. *IOSR Journal of Computer Engineering, 21*(3), 45–52.

Gyamfi, N. K., & Abdulai, J. D. (2018, November). Bank fraud detection using support vector machine. In *2018 IEEE 9th Annual Information Technology, Electronics and Mobile Communication Conference (IEMCON)* (pp. 37–41). IEEE.

Irvin-Erickson, Y., & Ricks, A. (2019). Identity theft and fraud victimization: What we know about identity theft and fraud victims from research-and practice-based evidence.

Itoo, F., & Singh, S. (2021). Comparison and analysis of logistic regression, Naïve Bayes and KNN machine learning algorithms for credit card fraud detection. *International Journal of Information Technology*, *13*(4), 1503–1511.

Khan, H., & Martin, K. M. (2020). A survey of subscription privacy on the 5G radio interface-The past, present and future. *Journal of Information Security and Applications*, *53*, 102537.

Khan, S., Chouhan, V., Chandra, B., & Goswami, S. (2014). Sustainable accounting reporting practices of Indian cement industry: An exploratory study. *Uncertain Supply Chain Management*, *2*(2), 61–72.

Kolli,CS. Devi,TU (2019). Isolation Forest and Xg, Boosting For Classifying Credit Card Fraudulent Transactions, *8*(8), 41–47.

Krishna Rao, N. V., Harika Devi, Y., Shalini, N., Harika, A., Divyavani, V., & Mangathayaru, N. (2021). Credit Card Fraud Detection Using Spark and Machine Learning Techniques. In *Machine Learning Technologies and Applications* (pp. 163–172). Springer, Singapore.

Kulkarni, A. Ghare P (2019). Credit Card Fraud Detection Using Isolation Forest and Local Outlier Factor. *Annals of the Romanian Society for Cell Biology*, 4391–4396.

Kumar, M. S. Soundarya, V. Kavitha, S. Keerthika E. S. and Aswini, E (2019). Credit Card Fraud Detection Using Random Forest Algorithm, 2019 3rd International Conference on Computing and Communications Technologies (ICCCT), 149–153

Kurien, K. L., & Chikkamannur, A. A. (2019, May). Benford's Law and Deep Learning Autoencoders: An approach for Fraud Detection of Credit card Transactions in Social Media. In *2019 4th International Conference on Recent Trends on Electronics, Information, Communication & Technology (RTEICT)* (pp. 1030–1035). IEEE.

Levi, M. (2008). Organized fraud and organizing frauds: Unpacking research on networks and organization. *Criminology & Criminal Justice*, *8*(4), 389–419.

Loginovsky, O. V., Shestakov, A. L., & Shinkarev, A. A. (2020). Supercomputing technologies as drive for development of enterprise information systems and digital economy. *Supercomputing Frontiers and Innovations*, *7*(1), 55–70.

Nesvijevskaia, A., Ouillade, S., Guilmin, P., & Zucker, J. D. (2021). The accuracy versus interpretability trade-off in fraud detection model. *Data & Policy*, *3*.

Nipane, V. B., Kalinge, P. S., Vidhate, D., War, K., & Deshpande, B. P. (2016). Fraudulent detection in credit card system using SVM & decision Tree. *International Journal of Scientific Development and Research (IDSDR)*, *1*(5), 590–594.

Nipane, V. B., Kalinge, P. S., Vidhate, D., War, K., & Deshpande, B. P. (2016). Fraudulent detection in credit card system using SVM & decision Tree. *International Journal of Scientific Development and Research (IDSDR)*, *1*(5), 590–594.

Niu, X., Wang, L., & Yang, X. (2019). A comparison study of credit card fraud detection: Supervised versus unsupervised. *arXiv preprint arXiv:1904.10604*.

Puh, M., & Brkić, L. (2019, May). Detecting credit card fraud using selected machine learning algorithms. In *2019 42nd International Convention on Information and Communication Technology, Electronics and Microelectronics (MIPRO)* (pp. 1250–1255). IEEE.

Quah, J. T., & Sriganesh, M. (2008). Real-time credit card fraud detection using computational intelligence. *Expert systems with applications*, *35*(4), 1721–1732.

Raj, P., & Raman, A. C. (2015). *Intelligent Cities: Enabling tools and technology*. CRC Press.

Raj, S. B. E., & Portia, A. A. (2011, March). Analysis on credit card fraud detection methods. In *2011 International Conference on Computer, Communication and Electrical Technology (ICCCET)* (pp. 152–156). IEEE.

Rajora, S., Li, D. L., Jha, C., Bharill, N., Patel, O. P., Joshi, S.,... & Prasad, M. (2018, November). A comparative study of machine learning techniques for credit card fraud detection based on time variance. In *2018 IEEE Symposium Series on Computational Intelligence (SSCI)* (pp. 1958–1963). IEEE.

Randhawa, K., Loo, C. K., Seera, M., Lim, C. P., & Nandi, A. K. (2018). Credit card fraud detection using AdaBoost and majority voting. *IEEE access*, *6*, 14277–14284.

Robertson, D (2016). Investment and & Acquisitions Top Card Issuers in Asia-Pacific Card Fraud Losses Reach $21.84 Billion, Nilson Rep, 1096-1090.

Rushin, G., Stancil, C., Sun, M., Adams, S., & Beling, P. (2017, April). Horse race analysis in credit card fraud—deep learning, logistic regression, and Gradient Boosted Tree. In *2017 systems and information engineering design symposium (SIEDS)* (pp. 117–121). IEEE.

Shen, A., Tong, R., & Deng, Y. (2007, June). Application of classification models on credit card fraud detection. In *2007 International conference on service systems and service management* (pp. 1–4). IEEE.

Singh, G., Gupta, R., Rastogi, A., Chandel, M. D., & Ahmad, R. (2012). A machine learning approach for detection of fraud based on svm. *International Journal of Scientific Engineering and Technology, 1*(3), 192–196.

Sudha, C., & Raj, T. N. (2017). Credit card fraud detection in internet using k-nearest neighbor algorithm. *International Journal of Computer Science, 5*(11), 22–30.

Thennakoon, A., Bhagyani, C., Premadasa, S., Mihiranga, S., & Kuruwitaarachchi, N. (2019, January). Real-time credit card fraud detection using machine learning. In *2019 9th International Conference on Cloud Computing, Data Science & Engineering (Confluence)* (pp. 488–493). IEEE.

Trivedi, N. K., Simaiya, S., Lilhore, U. K., & Sharma, S. K. (2020). An efficient credit card fraud detection model based on machine learning methods. *International Journal of Advanced Science and Technology, 29*(5), 3414–3424.

Wei, W., Li, J., Cao, L., Ou, Y., & Chen, J. (2013). Effective detection of sophisticated online banking fraud on extremely imbalanced data. *World Wide Web, 16*(4), 449–475.

Xuan, S., Liu, G., Li, Z., Zheng, L., Wang, S., & Jiang, C. (2018, March). Random forest for credit card fraud detection. In *2018 IEEE 15th international conference on networking, sensing and control (ICNSC)* (pp. 1–6). IEEE.

Zhang, X., Han, Y., Xu, W., & Wang, Q. (2021). HOBA: A novel feature engineering methodology for credit card fraud detection with a deep learning architecture. *Information Sciences, 557*, 302–316.

9

Integrating InSAR, GNSS, IoT, 5G, and Cybersecurity for Earthquakes/Tremor Monitoring and Forecasting in Abuja, Nigeria

Omega John Unogwu[1], Kamal Kant Hiran[2], Ruchi Doshi[3], and Manish Dadhich[4]

[1]Centre for Geodesy and Geodynamics, National Space Research and Development Agency, Nigeria and Universidad Azteca, Mexico
[2]Sir Padampat Singhanai University, India
[3]Universidad Azteca, Mexico
[4]Sir Padampat Singhanai University, India

Abstract

Application of radar interferometry to measure geophysical changes of the earth has rapidly evolved since the early 1990s. Interferometric Synthetic Aperture Radar (InSAR) is suitable for measurement of the spatial extent and magnitude of surface deformation associated with natural hazards (earthquakes, volcanoes, landslides, subsidence) and fluid extraction. The emergence of this new geodetic technique calculates the interference pattern that resulted from phase difference between two images of the same location, acquired by InSAR at different passes, producing a contour map or interferogram of the change in distance between the radar instrument and the ground. This records surface movement and change in the topography of the location. Integrating global navigation satellite system (GNSS) with established seismic monitors and 5G technology in Internet of Things – IoT (a platform where everyday devices become smarter, everyday processing becomes intelligent, and everyday communication becomes informative) for quakes and deformation monitoring is a novel idea that this study attempts to establish, with reference to the earth tremors of Abuja – Nigeria. The IoT architecture is capable of improving understanding of relevant technologies, tools, and methodologies at the most basic levels of the importance of InSAR for deformation monitoring through the following ways: (1) relevant data acquisition, (2) insight acquired from analyzing collected data, as well as (3) relevant actionable activities at the right place and time from the acquired insight. The approach of connected things facilitates acquisition of information that was not available previously, and expand their reach beyond traditional geographic boundaries. Devices linked to the Internet for transmission of InSAR data are subject to cyberattacks, which downgrades data integrity. Insecure protocols, nonencrypted data and poor user-authentication mechanisms are typical cybersecurity problems. The aggregate of timeliness, environmental characteristics, locational awareness as well as data-transmitting devices/ security permits a new level of operational efficiency which is a highly relevant factor in integrating InSAR, GNSS, 5G, IoT, and cybersecurity for earthquakes and deformation monitoring.

9.1 Introduction

Natural disasters are unsuspected catastrophic events. These disasters have been known to man since the stone ages. Landslides, subsidence, flooding as well as earthquakes are examples. They are known to result in destruction with varying degrees and magnitude in their wake. Human populations in different parts of the world are increasingly getting concerned, with researchers tirelessly engaging in the science behind these hazards and looking for novel ideas to mitigate their effects. According to (Chen et al., 2000), earthquakes occur as shakings, most of which are either earthquakes or tremors. These two are loosely termed as earthquakes. The shaking with higher degree and accompanied by severe damages are earthquakes, while those with minor shakings are termed tremors. An earthquake otherwise known as a tremor is a shaking of the surface of the earth as a result of the sudden release of energy in the earth crust (Nwankwoala & Orji, 2018). Tremors usually occur in a region as a sign that earthquakes of higher magnitude is likely to occur.

Earthquakes and other geological deformations are becoming increasingly, of great concern to all regions of the world. These geological deformations involve the geographical environment, geological conditions, large spatiotemporal differences in climate, and other hidden and widely distributed geological phenomena which are destructive and occur suddenly (King, 2015). As a result of these, coupled with human activities like constructions, mining and drillings over time, destructive geophysical disasters occur more frequently as have been recorded in recent times. According to (Guilhot et al.,

2021), mining activities such as excavation, blasting, material removal, and water extractions can trigger such events. Geohazards are generally induced by a series of external inducers and internal geological conditions (Jensen & Aven, 2017; Pontes et al., 2012). In the earlier years, monitoring of these geohazards was largely undertaken by manual observations of surface change characteristics due to equipment limitations (King, 2015). However, geohazard prevention/limitation has evolved over time due to advances and application of geological engineering (Abraham et al., 2020), application of space platforms like Interferometric Synthetic Aperture Radar (InSAR), as well as emerging technologies like IoT and 5G to minimize or prevent the effects of the various geologic hazards (Priyadarshi et al., n.d.). Geohazards monitoring systems based on the IoT (Hiran Kamal Kant. Doshi Ruchi., 2014) require universal, high-performance and scalable services to meet the requirements of real-time monitoring in complex and harsh geological environments. Combining networks of geotechnical surface sensors like GNSS and seismometers with InSAR data has proved to accurately monitor surface displacement (Guilhot et al., 2021). Monitoring and early warning are the most common strategies for geohazards prevention. Real-time monitoring and intelligent early warning system are crucial and significant to take mitigation measures and reduce casualties and property losses related to deformations (Xu et al., 2020). Early warning is primarily meant to give out alerts before hazardous occurrences take place to curtail impending disasters to life and property.

Nigeria was not considered to be situated on a significantly active seismic zone on the earth surface. However, historic data indicates pockets of earthquakes and tremors have been recorded over time spanning close to a century between 1933 till date (Guilhot et al., 2021; Hanssen, n.d.). More significantly and contrary to expectations, Abuja – Nigeria, thought to be on an aseismic belt of the country experienced a series of tremors with significant ground shaking (Ebi et al., 2021; Guilhot et al., 2021), which monitoring instruments recorded the stress build-up spanned three days prior to the actual event.

The location of urban zones at risk is crucially important for hazard mitigation policies as a result of ground instabilities of various origins (landslides, active faults, high deformable soils, underground salt deposits, and gas reservoirs) (Abraham et al., 2020; Lanari et al., 2004) which could be significantly prevalent.

This research aims to achieve the following objectives:

• To advance a self-adaptive data acquisition monitoring technique that will automatically show the displacement from InSAR as well as other

Table 9.1   Location of current and proposed seismic stations in Nigeria.

S#	Station code	Name	Coordinates	Geological foundation	Installed instruments
1	ABJ	Abuja	08°59'126"N 07°23'380"E	Granite	Non
2	ABK	Abakaliki	06°23'45"N 08°01'474"E	Sandstone	EP-105, Broadband seismometers, DR4000 recorder
3	AWK	Awka	06°14'561"N 07°06'693"E	Shale and siltstone	EP-105, Broadband seismometers, DR4000 recorder
4	IBN	Ibadan	07°27'251"N 03°53'520"E	Gneiss	Non
5	IFE	Ile-Ife	07°32'800"N 04°32'815"E	Gneiss	EP-105, Broadband seismometers, DR4000 recorder
6	KAD	Kaduna	10°26'101"N 07°38'484"E	Granite	EP-105, Broadband seismometers, DR4000 recorder
7	MINN	Minna	09°30'702"N 06°26411"E	Granite gneiss	EP-105, Broadband seismometers, DR4000 recorder
8	NSU	Nsukka	06°52'011"N 07°25'045"E	Sandstone	EP-105, Broadband seismometers, DR4000 recorder
9	OYO	Oyo	07°53'131"N 03°57'078"E	Granite	SP-400 Seismometer, DR4000 recorder
10	TORO	Toro	10°26'303"N 09°07'089"E	GNEISS	EP-105, Broadband seismometers, DR4000 recorder

in-situ measuring devices like GNSS and seismometers (Higgins, 2010; "Living on an Active Earth", 2003).

- To establish a real-time 5G-enabled IoT (Mei et al., 2020; Ray, 2018) infrastructure for tremor and deformation monitoring and

- To recommend a practical proactive early warning model for monitoring established tremor-prone locations for predictive purposes on the event of a future deformation occurrences. Mpape and Maitama districts in Abuja are selected as benchmarks for implementation of a real-time monitoring and smart early warning system for deformations.

9.2 Case Study

9.2.1 Background of the Abuja 2018 tremors

Abuja is located at the central region and serves as the administrative headquarters of Nigeria. The territory experienced two tremors within a short period of time in 2018 (three days interval). This calls for urgent attention on the city to unravel the immediate and remote causes of the tremors as well as extending monitoring activities beyond mere postmortem practices, to implementation of proactive monitoring for effective early warning mechanisms, keeping in mind that urban areas are very complex geological, geographical, physical and social environments, characterized by high vulnerabilities. Consequently, it is very difficult to perform reliable hazard assessments within such settings (K. U. Afegbua et al., 2018) based merely on geophysical monitoring installations.

Abuja has seen over 30 years of consistent constructions in varying degrees following the relocation of the administrative capital of Nigeria from Lagos. This, perhaps contribute to factors alongside its geology, to the occurrence of the tremors. Hence, there is urgent need to establish a threshold with every conceivable sensing technology deployment for early warning on the event of impending seismic activity.

The September/November 2018 tremors at Mpape and Maitama areas of Abuja (Figure 9.1) potentially mark Nigeria as a considerable environment for natural hazards. It is necessary to develop a new professional monitoring technique, which could wirelessly, securely transmit displacement data, automatically adjust the data acquisition frequency according to the evolution characteristics of stress build up within the crust and the rate increment of displacement in the city (Adepelumi, 2020).

Abuja Tremors
Area wey e happen

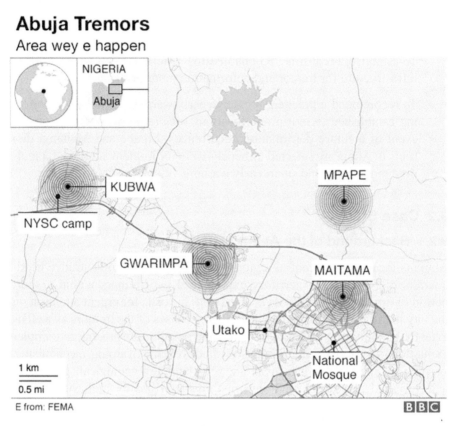

Figure 9.1   Locations of the 2018 tremors – Mpape and Maitama (Adepelumi, 2020).

9.3 Research Methodology

Numerous geophysical methods abound for earthquakes/deformation research that covers regional expanse. The methods adopted in this research utilizes the principles of InSAR, GNSS, IoT, 5G technology, and cybersecurity as a robust (Choubisa & Doshi, 1 C.E.; Easwaran et al., 2022), multifaceted package to serve as an effective monitoring, early warning as well as a reliable and accurate data acquisition/management system for near real-time to real-time geohazard monitoring – adopting the proactive approach.

9.3.1 InSAR

InSAR, a nested acronym for radio detection and ranging (RADAR (el Kamali et al., 2020; Musa et al., 2016)), with synthetic aperture radar (SAR) – a

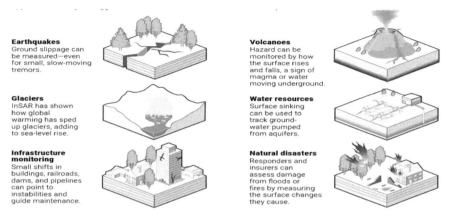

Earthquakes
Ground slippage can be measured—even for small, slow-moving tremors.

Glaciers
InSAR has shown how global warming has sped up glaciers, adding to sea-level rise.

Infrastructure monitoring
Small shifts in buildings, railroads, dams, and pipelines can point to instabilities and guide maintenance.

Volcanoes
Hazard can be monitored by how the surface rises and falls, a sign of magma or water moving underground.

Water resources
Surface sinking can be used to track ground-water pumped from aquifers.

Natural disasters
Responders and insurers can assess damage from floods or fires by measuring the surface changes they cause.

Figure 9.2 Diverse areas of InSAR applications. Source: https://www.science.org/content/article/fleets-radar-satellites-are-measuring-movements-earthnever

remote sensing technique using two or more SAR phase images acquired at different times to generate maps to detect and map changes of spatial and/or dielectric properties of the ground surface by using differences in the phase of the waves returning to the satellite or aircraft (Perski et al., 2009; Pritchard, 2006; Rocca et al., 2007). SAR data are usually delivered to users in frames (Figure 9.2) corresponding to an area of about 100×100 km². Therefore, its application on Abuja on a continuous basis is expedient to help define the established benchmarks for early warnings on the event of further quakes (Pritchard, 2006; Zhou et al., 2009).

InSAR (Massonnet et al., 1994; Tapete et al., 2013) can provide distinctive views of surface disruptions through measurements of interferometric decorrelation, which has potentials to aid emergency responders' ability to respond efficiently to many natural disasters. InSAR as a complimentary tool can operate in conjunction with the already established continuously operating ground measurement instruments like GPS (Ge et al., 2001) and seismometers to observe deformation in time scales in seconds.

That InSAR can take advantage of a satellite's perspective of the world permits one to view large areas of Earth's surface (Xu et al., 2020) quickly and efficiently due to the following benefits:

- High precision: ability to detect displacements of 1–2 mm per year

- Large spatial coverage: Images (areas of interest) in excess of 1500 km² each

- Dense coverage: Typically generates tons of data points covering areas of interest

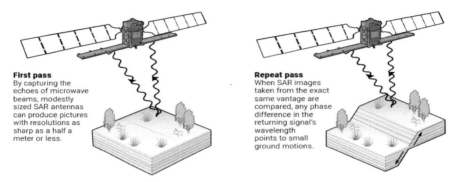

Figure 9.3　Simplified operational principle of SAR. Source: https://www.science.org/content/article/fleets-radar-satellites-are-measuring-movements-earthnever

- Remote sensing: No ground instruments or on-site inspection required

- Full-site monitoring: Often detects displacement measurement of unknown risks (Figure 9.3).

- Measurement frequency: Updated displacement measurements after each satellite revisit (basically 2–12 days)

9.3.2　5G and IoT Architecture

With the development of conventional technological instruments of surface displacement monitoring, including GNSS monuments – based on point data acquisition, and Interferometric Synthetic Aperture Radar (InSAR) technology, are deployed to monitor the primary locations of Mpape and Maitama (Adepelumi, 2020). This has resulted in great progress in monitoring and is currently adding to the prediction potentials of tremors. 5G communication and IoT technologies (Kovatsch, 2015; Li et al., 2021) are vital to establish an efficient remote-monitoring system of geological catastrophes. Compared with the older techniques, the 5G communication recently licensed for deployment in Nigeria, delivers a higher speed, greater capacity, and lesser latency, and the IoT (Tuly, 2016) permits interconnection of diverse sensors in one framework to synthetically analyze and assess a problem. In consideration of the abovementioned techniques, the remote-monitoring system architecture is applied. Relevant sensors are installed in the field to obtain necessary data (Agrawal & Vieira, 2013), which are remotely transmitted and stored in servers. After the analysis, according to the tremor benchmark criterion, the corresponding warning can be sent. With the rapid development

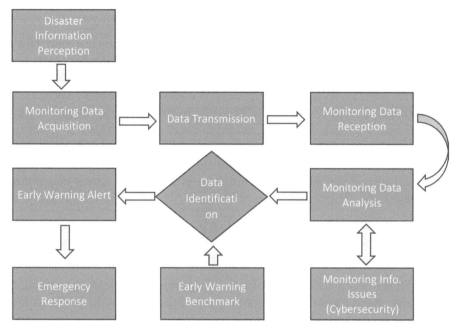

Figure 9.4 Process logic for earthquake/deformation with early warning system mechanism.

of 5G communication technology, a novel monitoring and prediction system based on 5G–IoT is under development to be applied in Abuja (K. Afegbua et al., 2018; Alaneme & Okotete, 2018).

This chapter describes the monitoring and early warning system based on 5G–IoT technology. The system monitors important indicators such as three-dimensional surface displacement, and ground cracks through GNSS equipment, with InSAR providing important data over the coverage area, and various IoT sensors deployed on site, transmitting key monitoring data with 5G communication technology for long-term monitoring and as early warning mechanism ("AN Internet of Things Based Software Defined Security Service Provisioning Framework", 2020; Tawalbeh et al., 2020). An early warning guidelines-based threshold is established in accordance with a four-level early warning method by combining the stress buildup rate per time as well as embedded vector maps such as topographic and geo-morphological remote sensing maps of Abuja, the geological section map, and the monitoring points distribution map. The system has the functions of accurate acquisition, rapid transmission, automatic search, and comprehensive analysis (Figure 9.4).

9.3 Results and Discussions

Thresholds of seismic activities within Abuja are being established and monitoring stations taking form with established benchmarks for early warning. The relationship between tremors and catastrophes varies from site to site due to the peculiarities of the morphologies under consideration. Therefore, this approach considered in this work needs to be customized based on the distinctiveness of each environment for the sake of a suitable early warning system.

Alert criteria are required to establish a connection between the magnitude of earth shakings in Abuja and a set of alert levels as set out in the following:

• Threshold level 1: Attention level. When the vibration rate exceeds the default threshold, it identifies the state of shaking begins to have abnormal tremors or enter the constant vibration phase.

• Threshold level 2: Caution level. When vibrations exceed the default value, it indicates that the abnormal tremor has entered the initial acceleration phase under normal circumstances.

• Threshold level 3: Vigilance level. When the vibrations exceed the default value, it means that the tremor has exceeded the state of short-term rapid vibrations or has entered the medium-term acceleration phase and imminent quake phase.

9.4 Conclusion

This research defines monitoring and early warning system for geohazards based on 5G communication networks side by side with the established GNSS monuments, and seismic ground stations, with InSAR technology as a space-borne platform for continual monitoring, because 5G requires high data encryption processing and security (Hiran et al., 2012).

The authors report the suitability of integrating these techniques, vis a vis, InSAR, GNSS, 5G, IoT, and cybersecurity architecture in Nigeria for effective coverage on geohazard-prone areas, thereby instituting an early warning system in terms of quakes, tremors, landslides, subsidence, and the likes, to keep ahead on the event of any geohazards. Furthermore, integration of these technologies clearly shows that GNSS can effectively show stress buildup within the crust which is a principal requirement for developing a benchmark for vibration characterization, while InSAR data measures millimeter-scale ground displacements over a wide expanse of ground on a large time scale, detecting minute displacements from crustal vibrations; and IoT architecture serves as the transmission medium of data and actual deformation events, while at the same time clearly shows that the in-situ sensors data are also used to optimize the InSAR data analysis for early warning activities.

The analysis of the applications, technologies, and challenges of the IoT in geohazards mitigation has been presented in this chapter. The appraisal of the literature reveals that the IoT is widely used in the mitigation of the effect of three common categories of geohazards: (1) earthquakes, (2) deformations, including surface cracks, surface collapse, and surface subsidence, and (3) slope failures, including landslides, debris flow, and rockfall. Several key technologies in IoT-based monitoring and early warning systems for geohazards mitigation have also been discussed in this chapter. Compared with the conventional human-based monitoring and early warning systems for geohazards mitigation, the IoT-based technology is more reliable for its accuracy, speed, safety, timeliness, and smart characteristics. However, there still are challenges to be addressed. First, security and unified standards should be seriously guaranteed. Second, in view of the complex and harsh environment of geohazards, the power consumption and sensor reliability should be enhanced more with the exploitation of renewable energy resources. Moreover, the utilization of emerging technologies such as deep learning, edge computing, and 5G in IoT-based monitoring and early warning systems for geohazards monitoring should be encouraged.

Finally, this research has focused on the security of data being communicated between highly constrained IoT devices and the Internet for the benefit of the society.

References

Abraham, E. M., Nkitnam, E. E., & Itumoh, O. E. (2020). Integrated geophysical investigation of recent earth tremors in Nigeria using aeromagnetic and gravity data. *Environmental Monitoring and Assessment, 192*(6). https://doi.org/10.1007/s10661-020-08339-6

Adepelumi, A. A. (2020). *Abuja Tremor: A Closer Look on Seismic Data Gathering in Nigeria a Paper Presented at the 2nd National Borehole Master Drillers Conference At The Transcorp Hilton Hotel, Abuja Hydrocarbon reservoir characterization and discrimination using well-logs over. January*. https://doi.org/10.13140/RG.2.2.14491.21284

Afegbua, K., Ezomo, F., Osahon, O., Kijko, A., Smit, A., & Dimas, V.-A. (2018). Tectonic Activities of the Mid-Atlantic Ridge and Implication of Seismicity in West African Region. *Journal of Scientific Research and Reports*, *18*(4). https://doi.org/10.9734/jsrr/2018/39760

Afegbua, K. U., Yakubu, T. A., Sanni, H. T., Oluwafemi, O., Karabulut, H., & Cambaz, M. D. (2018). A Preliminary Evaluation of Broadband Stations in Nigeria. *Advances in Research*, *17*(4). https://doi.org/10.9734/air/2018/39641

Agrawal, S., & Vieira, D. (2013). A survey on Internet of Things - http://dx.doi.org/10.1016/j.jksuci.2016.10.003. *Abakós*, *1*(2).

Alaneme, K. K., & Okotete, E. A. (2018). Critical evaluation of seismic activities in Africa and curtailment policies – a review. In *Geoenvironmental Disasters* (Vol. 5, Issue 1). https://doi.org/10.1186/s40677-018-0116-2

AN Internet of Things based Software Defined Security Service Provisioning Framework. (2020). *International Journal of Recent Technology and Engineering*, *8*(6), 3358–3362. https://doi.org/10.35940/ijrte.f8735.038620

Chen, Y., Zhang, G., Ding, X., & Li, Z. (2000). Monitoring Earth Surface Deformations with InSAR Technology: Principles and Some Critical Issues. *Journal of Geospatial ...*, *2*(July), 3–21. http://www.lsgi.polyu.edu.hk/staff/zl.li/vol_2_1/02_chen.pdf

Choubisa, M., & Doshi, R. (1 C.E.). Crop Protection Using Cyber Physical Systems and Machine Learning for Smart Agriculture. *Https://Services.Igi-Global.Com/Resolvedoi/Resolve.Aspx?Doi=10.4018/978-1-7998-9308-0.Ch009*, 134–147. https://doi.org/10.4018/978-1-7998-9308-0.CH009

Easwaran, B., Hiran, K. K., Krishnan, S., & Doshi, R. (Eds.). (2022). *Real-Time Applications of Machine Learning in Cyber-Physical Systems*. https://doi.org/10.4018/978-1-7998-9308-0

Ebi, S., Nwankwo, C. N., Isogun, A. M., Afegbua, U. K., & Ibitola, A. R. (2021). *Determination of Compressional To Shear Wave Velocity Ratio from Local Earthquakes in Nigeria between (2009–2018) Determination of Compressional To Shear Wave Velocity Ratio from Local Earthquakes in Nigeria between (2009–2018)*. *May*. https://doi.org/10.9790/0990-0802022837

el Kamali, M., Abuelgasim, A., Papoutsis, I., Loupasakis, C., & Kontoes, C. (2020). A reasoned bibliography on SAR interferometry applications and outlook on big interferometric data processing. In *Remote Sensing Applications: Society and Environment* (Vol. 19). https://doi.org/10.1016/j.rsase.2020.100358

Ge, L., Chen, H., Han, S., & Rizos, C. (2001). Integrated GPS And Interferometric SAR Techniques For Highly Dense Crustal Deformation Monitoring. *Data Processing*, *September*, 11–14.

Guilhot, D., del Hoyo, T. M., Bartoli, A., Ramakrishnan, P., Leemans, G., Houtepen, M., Salzer, J., Metzger, J. S., & Maknavicius, G. (2021). Internet-of-things-based geotechnical monitoring boosted by satellite insar data. *Remote Sensing, 13*(14). https://doi.org/10.3390/rs13142757

Hanssen, R. F. (n.d.). Chapter 5 - Data analysis and interpretation for deformation monitoring. *InSAR Interpretation Analysis.*

Higgins, N. (2010). Chapter 7. Conclusions And Recommendations. In *Regulating the Use of Force in Wars of National Liberation: The Need for a New Regime.* https://doi.org/10.1163/ej.9789004172876.i-272.59

Hiran, K. K., Jain, R. K., Hiran, K., & Paliwal, G. (2012). Quantum Cryptography: A New Generation of Information Security System Role of Internet Access Infrastructure on Traveler Behaviour in Intelligent Transportations Systems within the Smart City Concept View project Cloud Computing View project Quantum Cryptography: A New Generation of Information Security System. *International Journal of Computers and Distributed Systems Www.Ijcdsonline.Com, 2.* https://www.researchgate.net/publication/320404164

Hiran Kamal Kant. Doshi Ruchi. (2014). *Proliferation of Smart Devices on Mobile Cloud Computing.*

Jensen, A., & Aven, T. (2017). Hazard/threat identification: Using functional resonance analysis method in conjunction with the Anticipatory Failure Determination method. *Proceedings of the Institution of Mechanical Engineers, Part O: Journal of Risk and Reliability, 231*(4). https://doi.org/10.1177/1748006X17698067

King, J. (2015). *A Distributed Security Scheme to Secure Data Communication between Class-0 IoT Devices and the Internet.* 58.

Kovatsch, F. M. (2015). *Scalable Web Technology for the Internet of Things.* 22398.

Lanari, R., Lundgren, P., Manzo, M., & Casu, F. (2004). Satellite radar interferometry time series analysis of surface deformation for Los Angeles, California. *Geophysical Research Letters, 31*(23). https://doi.org/10.1029/2004GL021294

Li, Z., Fang, L., Sun, X., & Peng, W. (2021). 5G IoT-based geohazard monitoring and early warning system and its application. In *Eurasip Journal on Wireless Communications and Networking* (Vol. 2021, Issue 1). https://doi.org/10.1186/s13638-021-02033-y

Living on an Active Earth. (2003). In *Living on an Active Earth.* https://doi.org/10.17226/10493

Massonnet, D., Feigl, K., Rossi, M., & Adragna, F. (1994). Radar interferometric mapping of deformation in the year after the Landers earthquake. *Nature, 369*(6477). https://doi.org/10.1038/369227a0

Mei, G., Xu, N., Qin, J., Wang, B., & Qi, P. (2020). A Survey of Internet of Things (IoT) for Geohazard Prevention: Applications, Technologies, and Challenges. In *IEEE Internet of Things Journal* (Vol. 7, Issue 5). https://doi.org/10.1109/JIOT.2019.2952593

Musa, O. K., Kurowska, E. E., Schoeneich, K., Alagbe, S. A., & Ayok, J. (2016). Tectonic control on the distribution of onshore mud volcanoes in parts of the Upper Benue Trough, northeastern Nigeria. *Contemporary Trends in Geoscience*, *5*(1). https://doi.org/10.1515/ctg-2016-0003

Nwankwoala, H. O., & Orji, O. M. (2018). An Overview of Earthquakes and Tremors in Nigeria: Occurrences, Distributions and Implications for Monitoring. *International Journal of Geology and Earth Sciences*, *4*(4). https://doi.org/10.32937/ijges.4.4.2018.56-76

Perski, Z., Hanssen, R., Wojcik, A., & Wojciechowski, T. (2009). InSAR analyses of terrain deformation near the Wieliczka Salt Mine, Poland. *Engineering Geology*, *106*(1–2). https://doi.org/10.1016/j.enggeo. 2009.02.014

Pontes, E., A. Silva, A. A., E., A., & T., S. (2012). Earthquake Prediction: Analogy with Forecasting Models for Cyber Attacks in Internet and Computer Systems. In *Earthquake Research and Analysis - Statistical Studies, Observations and Planning*. https://doi.org/10.5772/23384

Pritchard, M. E. (2006). InSAR, a tool for measuring Earth's surface deformation. *Physics Today*, *59*(7). https://doi.org/10.1063/1.2337843

Priyadarshi, Neeraj., Padmanaban, Sanjeevikumar., Hiran, K. Kant., Holm-Nielson, J. Bo., & Bansal, R. C. (n.d.). *Artificial Intelligence and Internet of Things for Renewable Energy Systems*.

Ray, P. P. (2018). A survey on Internet of Things architectures. In *Journal of King Saud University - Computer and Information Sciences* (Vol. 30, Issue 3). https://doi.org/10.1016/j.jksuci.2016.10.003

Rocca, F., Ferretti, A., Monti-Guarnieri, A. v., Prati, C. M., & Massonnet, D. (2007). Part C InSAR processing: a mathematical approach. In *InSAR Principles: Guidelines for SAR Interferometry Processing and Interpretation*.

Tapete, D., Casagli, N., Luzi, G., Fanti, R., Gigli, G., & Leva, D. (2013). Integrating radar and laser-based remote sensing techniques for monitoring structural deformation of archaeological monuments. *Journal of Archaeological Science*, *40*(1). https://doi.org/10.1016/j. jas.2012.07.024

Tawalbeh, L., Muheidat, F., Tawalbeh, M., & Quwaider, M. (2020). applied sciences IoT Privacy and Security : Challenges and Solutions. *Mdpi*.

Tuly, K. F. (2016). *A Survey on Novel Services in Smart Home (Optimized for Smart Electricity Grid)*. 1–103.

Xu, Q., Peng, D., Zhang, S., Zhu, X., He, C., Qi, X., Zhao, K., Xiu, D., & Ju, N. (2020). Successful implementations of a real-time and intelligent early warning system for loess landslides on the Heifangtai terrace, China. *Engineering Geology, 278*. https://doi.org/10.1016/j.enggeo.2020.105817

Zhou, X., Chang, N. bin, & Li, S. (2009). Applications of SAR interferometry in earth and environmental science research. In *Sensors* (Vol. 9, Issue 3). https://doi.org/10.3390/s90301876

Index

About the Authors

 **Professor Darío M. Goussal** is a full-time professor, founding member and coordinator of the Rural Telecommunications Program of the School of Engineering-Universidad Nacional del Nordeste at Resistencia, Argentina (UNNE) where he has been teaching since 1980. He has authored about a hundred scientific publications and conducted research on strategic planning aspects of rural telecommunications, supply-demand studies, feasibility and long-term expansion behavior of small local networks, community telecenters, cognitive radios, white-space technologies, optical broadband and educational networks, waste electronic equipment recycling, and rural utility cooperatives. A specialist and Senior Expert of the International Telecommunications Union (ITU), he has participated in engineering, training, consulting, project evaluation and assessment missions for national government agencies, international organizations and consulting agencies in over 20 countries.

 **Gaurav Meena** has been an Assistant professor in the Department of Computer Science at the Central University of Rajasthan, Ajmer, since April 2013. Mr. Meena's expertise is in Machine Learning, Data Mining, and Information Security. He is an author of some books and published more than 25 papers in international conferences & reputed journals. Before joining the Central University of Rajasthan, he served as an Information Technology Manager in the industry and worked as a Software Engineer in Tata Consultancy Services. He is also a member of the Internet Society, IEEE & ACM.

Web: https://www.curaj.ac.in/faculty/gaurav-meena

Dr. Mehul Mahrishi, with a teaching and research experience of more than 13 years, Mr. Mahrishi is currently working as an Associate Professor and Dy. Head in the faculty of Information Technology at the Swami Keshvanand Institute of Technology, Management, and Gramothan, Jaipur – India. Mr. Mahrishi is a Senior Member of the IEEE-Delhi Section, a Life Member of the Institution of Engineers, India (L-IEI), and a Life Member of the International Association of Engineers (IAENG). Mr. Mahrishi was a member of an Indian contingent of BRICS International Forum that traveled to Moscow, Russia, to attend Russian Energy Week-2018 and BRICS Youth Energy Summit. He has published more than 20 research papers in national and international journals/conferences, including IEEE Access, Springer, Taylor & Francis.

Professor. Prashant. S. Dhotre has obtained his B.E. Degree in Computer Science and Engineering from Shri Ramanand Teertha Marathwada University, Nanded, India in 2004 and M.E. Degree in Information Technology from Savitribai Phule Pune University, Pune, India in 2010. He was awarded a Ph.D. degree in Computer Engineering by Aalborg University, Denmark in 2017. Mr. Dhotre's Ph.D. work has been appreciated by Denmark University and published as a news article titled "Plugin from Aalborg University warns about the invisible collection and sharing of user data" on version2.dk (link: https://www.version2.dk/artikel/plugin-aalborg-universitet-advarer-usynlig-indsamlingdeling-bruger-data-1084557). He has authored seven books with Springer and National Publishers. Mr. Dhotre has delivered several talks on "how to write research papers, Use of reference manager and citation tools", "How to complete the PhD within a time", "How to write a thesis: Tips and tricks", etc., for Ph.D. scholars and Ph.D. aspirants at university level. His research interests include Algorithms, Privacy, and Turing Machine. He has visited a few countries like Denmark, Sweden, China, and Germany. He has more than 17 years of teaching and research experience. Currently, he is working a Associate Professor in the Department of Information Technology of MIT School of Engineering, MIT Arts, Design, and Technology University, Pune, India.

Professor. Mohd Shafi Pathan is currently working as a Professor with the Department of Computer Science and Engineering, MITSOE, MIT School of Engineering, MIT Arts, Design, and Technology University, Pune. He has worked as a Lecturer with the MIT Engineering College, Aurangabad, from July 1999 to July 2006, and as a Lecturer and an Associate Professor with the Smt. Kashibai Navale College of Engineering, Pune, from July 2006 to January 2017. He completed the university funded research project on "Public key cryptography for cross-realm authentication in Kerberos" costing two lakh rupees within a duration of two years. Mr. Pathan has six Book Chapters, nine Books, 49 Research paper, three patents, and three Copyrights on his credit.

Dr. Nilesh P. Sable has overall 14 years of experience, presently working as SPPU Approved Associate Professor in the Department of Information Technology, Vishwakarma Institute of Information Technology, Pune, India. He has done a Ph.D. in Computer Science & Engineering from Kalinga University, Raipur on Research Problem Statement "Study on Relationship Standard Mining Calculations in Data Mining" – Ph. D awarded on 3rd June 2018. He obtained his M.Tech. (Information Technology) degree from JNTU, Hyderabad in 2014 and a B.E. (Information Technology) degree from the University of Pune, Pune in 2008. He is an SPPU-approved Ph.D. research guide. He has published 40+ papers in National, International conferences and journals. He had filed and published 10+ Patents and Copyrights. He is the author of a couple of books with an international publisher like Lambart.

Dr. Roslyn Layton is a Visiting Researcher at the Center for Communication, Media and Information Technologies (CMI), Aalborg University, Copenhagen. She earned a doctorate at CMI in 2017. She serves as Senior Vice President of Strand Consult, an independent consultancy performing research on the global mobile telecom industry, and co-founder of China Tech Threat, a think tank researching Chinese government technology.

Henoch Kwabena-Adade is a Network and Systems Integration Engineer. He is a former Network Infrastructure Administrator at the United Nations Emergency Unit, a Senior Network Security Consulting Engineer for Gold Fields Ghana, and Head of Technical at Redmango Ghana Ltd. Mr. Kwabena-Adade has successfully designed and implemented ICT projects for organizations including Tullow, Vodafone, Ministry of Finance, Ecobank, GIPC, GridCo, Gold Fields Ghana, and Tema Meridian Port Services to name a few. He holds a Master's Degree in Information Technology, a Bachelor's Degree in Computer Science, and HND in Mechanical Engineering.

Prof. Ezer Osei Yeboah-Boateng is a professional Telecoms Engineer and ICT Specialist with a strong hands-on expertise in a wide range of telecommunications switching systems, revenue assurance, cybersecurity, digital forensics, business development, digital transformation, project management, change management, knowledge management, and strategic IT-enabled business value creation and capabilities to develop market-oriented strategies aimed at promoting growth and market share. Mr. Yeboah-Boateng has recently been appointed as the Deputy Director General, Technical Operations, of the National Communications Authority (NCA), where he brings his rich experiences to promote the World-class communications regulator's vision and mandate. He is an executive with over 25 years of domestic and global experience conceptualizing ideas, seizing opportunities, building operations, leading highly successful new business development initiatives and ventures. Mr. Yeboah-Boateng served as a Team Leader and Lead Researcher on various projects in Telecommunications infrastructure, Cybersecurity, Cloud Computing, Digital Forensics, and Internet and network integration technologies.

Grace Dzifa Kwabena-Adade is currently an IT Engineer at Cenpower Operations & Services Limited, one of the largest Independent Power Plants in Ghana. She is a former Systems Support Engineer at Eni Ghana Exploration and Production Limited. Ms. Kwabena-Adade graduated from the Kwame Nkrumah University of Science and Technology with a bachelor's degree in Computer Science and holds MSc. in Management Information Systems from Coventry University.

Dr. Somayeh Labafi earned her master's and doctorate degrees in media management from the University of Tehran. Her master's and doctoral theses both focused on media policy.

Ms. Labafi began her academic career as a faculty member of the Iranian Research Institute for Information Science and Technology (IranDoc), and she is now an Assistant Professor in the Department of Society and Information Studies. Her research is mostly concerned with social media and media policy. Her research areas of interest include social media, media policy, and network analysis methods. Ms. Labafi is the director of the "Social Media and Data Governance" Laboratory at the moment. She welcomes international collaboration and participation in projects.

Email: Labafi@irandoc.ac.ir.

Ali Darvishi is a student of Media Management in Allameh Tabataba'i University. His research area of interest is social media. Email: darvishi@gmail.com

Hadi Moghadamzadeh is a Ph.D. Student of Human Resource Management in University of Tehran. His research area of interest is Human Resource Management. Moghaddam.h@ut.ac.ir

Robert Mwiinga is pursuing Ph.D. from Sir Padampat Singhania University, Udaipur. His area of interest is Economics, Finance, Sustainable Green Practice, and Entrepreneurship. Apart from his exposure to corporate, he has credits to several national and international conferences and research papers published in national and international journals.

Dr. Manish Dadhich has Ph.D. from the Department of Commerce, EAFM, University of Rajasthan; M.Com, UGC-NET (Commerce); MBA-FM, UGC-NET (Management), RPSC-SET (Management). He has 14+ years of teaching experience in various colleges, universities, and corporate sectors, a rare blend of academia, industry, corporate consultancy, and research. Mr. Dadich is presently working as Assistant Professor in School of Management, Sir Padampat Singhania University, Udaipur. He has published more than 52 research papers in reputed international & national journals, including 15 papers in Scopus/ SCI/IEEE/WoS. He also presented more than 50 research papers at national and international conferences. He is a reviewer, advisor, and editorial board member of various reputed International Journals and Conferences of IEEE, Springer, and Elsevier. Mr. Dadich was awarded two gold medals in National Seminar for the best research paper. He authored one edited book. He is a regular invitee for FDP, research workshops, orientation, and refresher course lectures in research. He is also awarded one Australian patent. Further, his main research work focuses on Finance, AI-ML, Economics, and Statistics.

Dr. Shubham Goswami is working as Assistant Professor, with the School of Management at Sir Padampat Singhania University, Udaipur (India). He holds a doctoral degree in Computer Science and Masters in Business Administration. He has been conducting workshops and Faculty Development Program on problem solving with management science. Mr. Goswami's area of research includes technology adoption, information systems, ICT for development, business analytics, and online marketing. He has presented his research work in several national and international conferences and published research papers in journals of national and international repute.

Dr. Bhawna Hinger is working as Assistant Professor in the Department of Accountancy and Business Statistics, Govt. M. G. College, Udaipur (Raj.). She has a Masters in Commerce and Ph.D with teaching experience of 14 years in post graduate courses.

Dr. Vineet Chouhan is working as Assistant Professor in the School of Management at Sir Padampat Singhania University, Udaipur Rajasthan. He is specialised in Financial, Cost and Management Accounting. He is having more than 18 years of experience in academics. Mr. Chouhan has obtained his Ph.D. degree from Mohanlal Sukhadia University, Udaipur. Seven students have been awarded their Ph.D. under his supervision. He has also qualified for UGC NET in Commerce and Management both and RPSC SLET-2001. Mr. Chouhan has published more than 90 research papers in journals of repute and attended more than 75 National and International Conferences/Seminars. He has written 14 books at UG, PG, and Reference books levels. He has also completed UGC, ICSSR, RUSA, and industry-sponsored consulting projects.

Dr. Tarannum Hussain is working as Post-doctoral Fellow in Management with Faculty of Management Studies, Mohanlal Sukhadia University, Udaipur. She has a Ph.D in Management, MBA in Marketing-Finance, and M.Com in Business Administration. She has also qualified NET-JRF in Management and Commerce. She had worked as Assistant Professor in Center for Agribusiness Management and Center for Technology Management at Maharana Pratap University of Agriculture and Technology in Udaipur.

Omega John Unogwu obtained a BSc and Education in Geography from the Ahmadu Bello University – Nigeria, an NIIT certified Network Engineer from NIIT Abuja – Nigeria, an MSc in Information Technology from Sikkim Manipal University, Accra – Ghana with numerous intermediary courses and research programs. He holds membership in elite professional organizations like the Institute of Electrical and Electronics Engineers (IEEE), The Nigerian Computer Society, The Nigerian Environmental Society, National Association of Geodesy among others.

Mr. Unogwu is a Chief Scientist at the National Space Research and Development Agency (NASRDA) – Nigeria, with over 13 years of scientific research experience in Space Science and Technology in the fields of Global

Navigation Satellite Systems (GNSS), Interferometric Synthetic Aperture Radar (InSAR), and Computing. He is currently a Ph.D. student in Computer Science at Azteca University, Mexico.

Kamal Kant Hiran is having more than 17 years of academic and research experience in Asia, Africa, Europe, and North America. He has published 15 International books with reputed publishers like BPB Publications, IGI Global, De Gruyter from India, Germany, USA, and UK. His book titled "Cloud Computing: Master the Concepts, Architecture and Applications with Real-World Examples and Case Studies" is popular among Indian and overseas market and got the best seller category in the emerging technology trends books. Currently, he is working as Adjunct Associate Professor in India, Denmark, and Mexico Country.

Mr. Hiran has made significant contributions to our society's technological transformation. He has several awards to his credit such as international travel grant for attending the IEEE Region 8 Committee Meeting at Warsaw, Poland; international travel grant for Germany from ITS Europe, Passau, Germany; Best Research Paper Award at the University of Gondar, Ethiopia; IEEE Liberia Subsection Founder Award; Gold Medal Award in MTech (Hons.); IEEE Ghana Section Award-Technical and Professional Activity Chair; IEEE Senior Member Recognition, Best IEEE Student Branch Award, Elsevier Reviewer Recognition Award. He has published 45 scientific research papers in SCI/Scopus/Web of Science, and IEEE Transactions Journal, conferences, two Indian Patents, one Australian Patent Grants. Mr. Hiran has made several international visits to Denmark, Sweden, Germany, Norway, Ghana, Liberia, Ethiopia, Russia, Dubai, Mexico, and Jordan for research exposures. His research interests focus on cloud computing, machine-deep learning, and intelligent IoT. His research work is appreciated by several media and renowned bodies across the world – ITS Europe, IEEE USA, Springer, Rajasthan Patrika, and Dainik Bhaskar.

Dr. Ruchi Doshi is having more than 15 years of academic, research, and software development experience in Asia, Europe, and Africa. Currently, she is working as a Research Associate Professor at the Universidad Azteca, Mexico, and Adjunct Professor at the Jyoti Vidyapeeth Women's University, Jaipur, Rajasthan, India. She worked in the BlueCrest University College, Liberia, West Africa; BlueCrest University College, Ghana, Africa; Amity University, Rajasthan, India; Trimax IT Infrastructure & Services, Udaipur, India.

Ms. Doshi is interested in the field of Machine Learning and Cloud computing framework development. She has published 20 research papers in peer-reviewed international journals and conferences; three Indian patents; three books on cloud computing, machine learning, mobile cloud computing, intelligent IoT systems for Big Data Analysis and mobile application development. She is the Reviewer, Advisor, Ambassador, and Editorial board member of various reputed International Journals and Conferences with IEEE, Springer, and Elsevier. She is an active member in organizing many international seminars, workshops, and conferences. She is nominated from the IEEE Headquarter, USA for the Chair, Women in Engineering (WIE) position in Liberia, West Africa.

About the Editors

Professor Knud Erik Skouby is professor emeritus, Aalborg University. Founding Director of Center for Communication, Media and Information Technologies, Aalborg University-Copenhagen (2007–2017). Mr. Skouby has a career as a university teacher and within consultancy since 1972; focus on ICT since 1987. Working areas: *Techno-economic Analyses; Development of Mobile/Wireless Applications and Services: Regulation of Telecommunications*

Mr. Skouby had worked as Project manager and partner in a number of international, European and Danish research projects. He had served on public committees within telecom, IT, and broadcasting: as member of boards of professional societies; as member of organizing boards, evaluation committees; and as invited speaker on international conferences. He published a number of Danish and international articles, books, and conference proceedings. He was a Member of EU's Economic and Social Council (1994–1998). He acted as past department chair of IEEE Denmark. Editor in Chief of Nordic and Baltic Journal of Information and Communication Technologies (NBICT); Chair of WGA in Wireless World Research Forum.

Dr. Prashant. S. Dhotre is a Associate Professor in the Department of Information Technology of MIT School of Engineering, MIT Arts, Design, and Technology University, Pune, India. He has obtained his B.E. Degree in Computer Science and Engineering from Shri Ramanand Teertha Marathwada University, Nanded, India in 2004 and M.E. Degree in Information Technology from Savitribai Phule Pune University, Pune, India in 2010. He was awarded a Ph.D. degree in Computer Engineering by Aalborg University, Denmark in 2017. His Ph.D. work has been appreciated by Denmark University and published as a news article titled "Plugin from Aalborg University warns about the invisible collection and sharing of user data" on version2.dk (link: https://www.version2.dk/artikel/plugin-aalborg-universitet-advarer-usynlig-indsamling-deling-brugerdata-1084557).

Mr. Dhotre has authored four books on a subject "Theory of Computations" and a book on "Machine Learning" for UG students of Computer Engineering. Recently, he has authored **two** books on "Context Aware Pervasive Systems and Applications" and "Internet of Things Integrated Augmented Reality" with Springer Publishers for researchers at International level. He is an Editor of two upcoming books named "Data Science for Civil Engineering: A beginner's Guide (CRC Press)" and "5G Security and Privacy in Developing Countries (River Publications)". He is also working as the Single Point of Contact for International Collaboration between Indian Education Institutes and Aalborg University, Denmark.

Mr. Dhotre loves to interact with the students through a series of guest lectures on "Design and Analysis of Algorithms", "Theory of Computation", and "Data Structures" in various engineering colleges across the Pune region.

As a part of his research, he has delivered several talks on "how to write research papers, Use of reference manager and citation tools", "How to complete the Ph.D within a time", "How to write a thesis: Tips and tricks", etc., for Ph.D. scholars and Ph.D. aspirants at university level. He has more than 18 years of teaching and research experience. He has guided more than 70 undergraduate students for projects. His publication includes more than 35 papers in the international conferences and journals. His research interests include algorithms, privacy, and Turing machine. He has visited a few countries like Denmark, Sweden, China, Germany, and UAE (Dubai).

Dr. Idongesit Williams is an Assistant professor at Aalborg University, Copenhagen. He manages and works on EU-funded projects and teaches master's degree courses on Green ICT and Internet Governance and economics. His areas of research are in ICT for development and digital transformation. He possesses a multi-disciplinary background and works primarily on interdisciplinary projects and initiatives focusing more into socio-economic, socio-technical related to information and communications technologies as well as the roll of the integrator. Mr. Williams holds a Ph.D. (obtained in 2015) and a master's in Information and Communication Technologies from Aalborg University, Denmark. He holds a bachelor's degree in physics. Research topics he has worked on are not limited to the facilitation of telecom and ICT infrastructure using Public Private Partnerships, the development and the sustenance of community-based networks, e-government implementation, science and technology studies, gender adoption of ICTs, organizational adoption of ICTs, ICT and migrant entrepreneurship, user experience with ICTs and organizational learning. Currently his interest is in the social cybersecurity and privacy implications to the use of ICTs in society. He has authored more than 70 research publications, including journal papers, books, book chapters, conference papers, and magazine articles.

Kamal Kant Hiran has more than 17 years of academic and research experience in Asia, Africa, Europe, and North America. He has published 15 International books with reputed publishers like BPB Publications, IGI Global, De Gruyter from India, Germany, USA, UK. His book titled "Cloud Computing: Master the Concepts, Architecture and Applications with Real-World Examples and Case Studies" is popular among Indian and overseas market and got the best seller category in the emerging technology trends books. Currently, he is working as Adjunct Associate Professor in India, Denmark, and Mexico Country.

Mr. Hiran has made significant contributions to our society's technological transformation. He has several awards to his credit such as International travel grant for attending the IEEE Region 8 Committee Meeting at Warsaw, Poland; International travel grant for Germany from ITS Europe, Passau, Germany; Best Research Paper Award at the University of Gondar, Ethiopia; IEEE Liberia Subsection Founder Award; Gold Medal Award in M.Tech (Hons.); IEEE Ghana Section Award – Technical and Professional Activity Chair; IEEE Senior Member Recognition, Best IEEE Student Branch Award, Elsevier Reviewer Recognition Award. Mr. Hiran has published 45 scientific research papers in SCI/Scopus/Web of Science and IEEE Transactions Journal, conferences, two Indian Patents, one Australian Patent Grants. He has made several international visits to Denmark, Sweden, Germany, Norway, Ghana, Liberia, Ethiopia, Russia, Dubai, Mexico, and Jordan for research exposures. His research interests focus on cloud computing, machine-deep learning, and intelligent IoT. His research work is appreciated by several media and renowned bodies across the world – ITS Europe, IEEE USA, Springer, Rajasthan Patrika, and Dainik Bhaskar.